Global Stability of Dynamical Systems

Michael Shub

Global Stability of Dynamical Systems

With the Collaboration of
Albert Fathi and Remi Langevin

Translated by Joseph Christy

With 43 Illustrations

Springer-Verlag
New York Berlin Heidelberg
London Paris Tokyo

Michael Shub
IBM
Thomas J. Watson Research Center
Yorktown Heights, NY 10598
U.S.A.

With the collaboration of Albert Fathi and Remi Langevin

Translated by Joseph Christy

AMS Subject Classifications: 34C35, 58F10, 54H20

ISBN 978-1-4419-3079-8

Original edition: *Stabilité globale des systèmes dynamiques* (Astérisque, Vol. 56). Société Mathématique de France, 1978.

Library of Congress Cataloging in Publication Data
Shub, Michael
Global stability of dynamical systems.
Translation of: Stabilité globale des systèmes dynamiques.
Includes bibliographies.
1. Topological dynamics. 2. Stability. I. Title.
QA611.5.S49713 1986 514 86-10164

To Alex and Steve

Contents

Introduction

These notes are the result of a course in dynamical systems given at Orsay during the 1976–77 academic year. I had given a similar course at the Graduate Center of the City University of New York the previous year and came to France equipped with the class notes of two of my students there, Carol Hurwitz and Michael Maller. My goal was to present Smale's Ω-Stability Theorem as completely and compactly as possible and in such a way that the students would have easy access to the literature. I was not confident that I could do all this in lectures in French, so I decided to distribute lecture notes. I wrote these notes in English and Remi Langevin translated them into French. His work involved much more than translation. He consistently corrected for style, clarity, and accuracy. Albert Fathi got involved in reading the manuscript. His role quickly expanded to extensive rewriting and writing. Fathi wrote (5.1) and (5.2) and rewrote Theorem 7.8 when I was in despair of ever getting it right with all the details. He kept me honest at all points and played a large role in the final form of the manuscript. He also did the main work in getting the manuscript ready when I had left France and Langevin was unfortunately unavailable. I ran out of steam by the time it came to Chapter 10. M. Yoccoz wrote this chapter based on a preliminary version of class notes by M. Lebasque.

The original English notes have disappeared. Anyway much writing and revising was done directly in French. Joe Christy undertook to translate the French book into English. I have used the opportunity to correct some small errors and sloppiness that made certain points of the French original seem "extraterrestrial." I have also added some appendices concerning the C^r center, center unstable, and strong unstable manifolds. These theorems have gained currency in recent years, but it is not easy to find clear presentations of them in the literature. The stable manifold theorem was written to make generalization easy and these new appendices are written with references back to the chapters on stable manifolds. Indeed, the entire approach is taken from Hirsch, Pugh, and Shub, *Invariant Manifolds*, Springer Lecture Notes, No. 583, and lends itself easily to the theory of normally hyperbolic invariant manifolds as carried out there.

The chapters are meant to be read in order except that the appendices and Chapter 9 may be omitted. There is a commentary and a bibliography for each chapter. These commentaries are subjective—they reflect how I learned the subject, and do not pretend in any way to provide a historical account of its development. Certain notions may be introduced in the commentary of some chapter which will only be justified in the commentary of a later chapter. Reference [*i.j*] is the *j*th reference in the bibliography of Chapter *i*. The chapters themselves are narrowly focused on the mathematical results. The commentaries sometimes give some perspective in which to view these results and should be read together with, or just after, the corresponding chapter.

Translator's Note

I would like to thank, first of all, Mary Ann Finnerty for typing the manuscript and devoting countless hours to typesetting and producing this book. My debt to her is immeasurable, as it is to Mike Shub; to them my gratitude is boundless. I should also like to thank Kristen Harber for the lovely pictures of the solenoid and Paul Blanchard, Pat McSwiggen, Alec Norton, and José Pedrozo for their help in proofreading.

CHAPTER 1

Generalities

Let M be a smooth compact manifold, and $f\colon M \to M$ a diffeomorphism. Our goal is to study the structure of the orbits of f, where the orbit of a point x is the set $\{f^n(x) | n \in Z\}$. We seek to describe the "history" of points of M; i.e., to follow them as we repeat f many times. First, the points with the simplest histories.

A *fixed point* of f is a point x such that $f(x) = x$; a *periodic point* is a fixed point for an iterate of f; that is, there exists a strictly positive n such that $f^n(x) = x$.

We now study some notions of recurrence which are weaker than periodicity; just as with periodicity they will make sense in the larger category of topological spaces and continuous mappings.

Definition 1.0. Let X be a topological space and $f\colon X \to X$ a continuous map. A point x of X is *wandering* if it has a neighborhood U such that $f^k(U) \cap U$ is empty for all positive k. A point x is *nonwandering* if the above does not hold; that is, if for all neighborhoods U of x, there is a positive k such that $f^k(U) \cap U$ is nonempty. We denote by $\Omega(f)$ the set of nonwandering points; when there is no fear of confusion we simply write Ω for this set.

Proposition 1.1. *Ω is closed and $f(\Omega)$ is contained in Ω. Moreover, if f is a homeomorphism, $f(\Omega) = \Omega$ and a point is nonwandering for f if and only if it is nonwandering for f^{-1}.*

PROOF. By definition the set of wandering points is open; thus its complement, Ω, is closed.

If x belongs to Ω and U is a neighborhood of $f(x)$, then $f^{-1}(U)$ is a neighborhood of x. Therefore, there is a k such that $f^k(f^{-1}(U)) \cap f^{-1}(U) \neq 0$;

the image of this intersection under f is contained in $f^k(U) \cap U$, which is thus nonempty.

If f is a homeomorphism and x is in $\Omega(f)$, then for every neighborhood U of x, there is a positive k such that $f^k(U) \cap U \neq 0$. The f^{-k} image of this intersection is contained in $U \cap f^{-k}(U)$, which is nonempty, so x is in $\Omega(f^{-1})$.

Finally, if f is a homeomorphism:

$$\Omega = f(f^{-1}(\Omega)) \subset f(\Omega) \subset \Omega, \qquad \text{so} \quad f(\Omega) = \Omega. \qquad \square$$

Since Ω is closed and a periodic point is clearly nonwandering, the closure of the set of periodic points, denoted $\text{Per}(f)$, is contained in the nonwandering set: $\overline{\text{Per}(f)} \subset \Omega(f)$.

Definition 1.2. Let $f\colon X \to X$ be a continuous map and x an element of X. The *ω-limit set* of x, $\omega_f(x)$, is defined by

$$\omega_f(x) = \{y \in X \mid \exists \text{ a sequence } n_i \to +\infty \text{ such that } f^{n_i}(x) \to y\}.$$

If f is a homeomorphism, we define the *α-limit set* of x similarly:

$$\alpha_f(x) = \{y \in X \mid \exists \text{ a sequence } n_i \to -\infty \text{ such that } f^{n_i}(x) \to y\}.$$

Further, we set $L_+(f) = \overline{\bigcup_{x \in X} \omega_f(x)}$, $L_-(f) = \overline{\bigcup_{x \in X} \alpha_x(x)}$ and $L(f) = L_+(f) \cup L_-(f)$.

Proposition 1.3. *$L(f)$ is contained in $\Omega(f)$.*

Proof. Let y be a point of $\omega(f)$ and U a neighborhood of y. Then there are integers $m > n > 0$ such that $f^m(x)$ and $f^n(x)$ are in U. Therefore, $f^{m-n}(U) \cap U$ is nonempty so y is nonwandering. A similar argument applies to the set of α limit points and f^{-1}. $\square$

Corollary 1.4. *Suppose X is sequentially compact and let U be a neighborhood of $\Omega(f)$. Then for any x in X, there is an $N > 0$ such that for all $n \geq N$, $f^n(x)$ belongs to U.*

Proof. Otherwise there would be ω limit points of x outside of U, away from $L_+(f)$. $\square$

Definition 1.5. Suppose X is endowed with a metric d. Given $f\colon X \to X$ and a positive real number ε, a sequence $\underline{x} = \{x_i \mid p < i < q;\ -\infty \leq p \leq \dot{q} - 2 \leq \infty\}$ is an *ε-pseudo-orbit* for f if $d(f(x_i), x_{i+1}) < \varepsilon$ for $p < i < q - 1$. An ε-pseudo-orbit x is *ε-pseudoperiodic* if there is an n, $0 < n \leq q - p - 2$, such that $x_i = x_{i+n}$ for all i with both x_i and x_{i+n} in $\underline{x}$. We say that a point x in X is ε-pseudoperiodic if it is the first term of a finite ε-pseudoperiodic ε-pseudo-orbit.

Intuitively ε-pseudo-orbits are sequences of points which one would take to be orbits if the positions of points were only known with a finite accuracy ε.

Definition 1.6. Let X be a metric space and $f: X \to X$. A point x in X is *chain recurrent* if it is ε-pseudoperiodic for all positive ε. We denote by $R(f)$ or even R the set of chain-recurrent points.

We leave it as an exercise for the reader to show that $R(f)$ is closed.

Proposition 1.7. *Let X be a metric space and $f: X \to X$ a continuous map. Then $\Omega(f)$ is contained in $R(f)$.*

PROOF. Let x be in Ω and ε be positive. We will show x is ε-pseudoperiodic. Choose $\delta < \varepsilon/2$ such that

$$d(x, y) < \delta \Rightarrow d(f(x), f(y)) < \frac{\varepsilon}{2}.$$

Let U be a neighborhood of x contained in the ball of radius δ about x. Since x is nonwandering, we can find an n such that $f^n(U) \cap U \neq 0$. If $n = 1$, $\{x, x\}$ is an ε-pseudoperiodic orbit; if $n > 1$, we can find a y in U with $f^n(y)$ in U and thus $\{x, f(y), \ldots, f^{n-1}(y), x\}$ is an ε-pseudoperiodic orbit. □

To recapitulate

$$\mathrm{Per}(f) \subset L(f) \subset \Omega(f) \subset R(f),$$

and it is a good exercise to find various f's where the inclusions are strict (see Figure 1.1)

One might ask if the inclusions are strict for "most" mappings, f. One way to make this precise is to put a topology on the space of diffeomorphisms so that we can talk about residual sets and generic properties.

If X is a compact metric space, we can put two metrics on $\mathrm{Hom}(X)$, the set of homeomorphisms of X onto X:

$$d_0(f, g) = \sup_{x \in X} d(f(x), g(x)),$$

$$d_C(f, g) = \sup_{x \in X} [d(f(x), g(x)), d(f^{-1}(x), g^{-1}(x))].$$

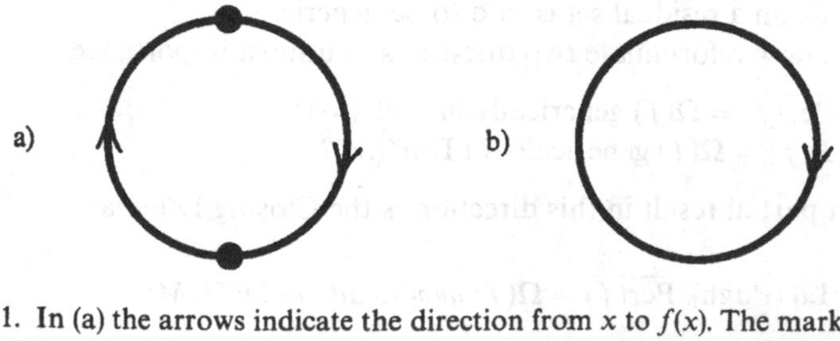

Figure 1.1. In (a) the arrows indicate the direction from x to $f(x)$. The marked points are fixed. Here $\mathrm{Per}(f) = L(f) = \Omega(f) =$ the two marked points, while $R(f)$ is the entire circle. In (b) f is a rotation through an irrational angle. $\mathrm{Per}(f)$ is empty, but $L(f) = \Omega(f) = R(f) =$ the entire circle.

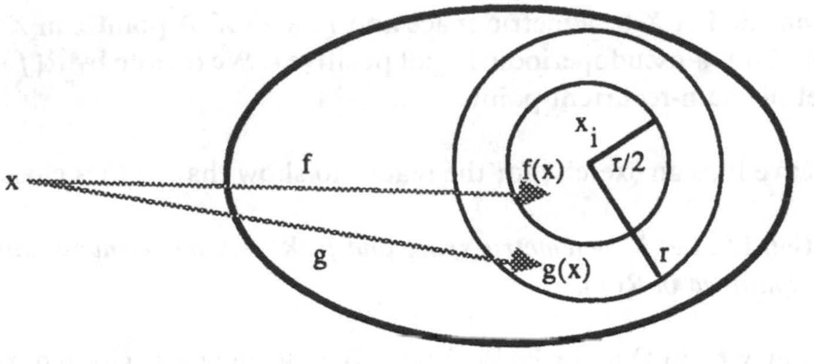

Figure 1.2.

These two metrics generate the same topology, since by compactness, all the mappings in question are uniformly continuous. Hom(X) is complete with the metric d_C.

If M is a compact C^r manifold, $1 \leq r < \infty$, we give the space of C^r diffeomorphisms, Diff$^r(M)$, a topology by specifying the convergent sequences: a sequence f_n converges C^r when all the derivatives of f_n of order less than or equal to r converge uniformly. This definition extends to the case $r = 0$ by putting Diff$^0(M) =$ Hom(M) as above.

The most natural way to make this definition precise is by introducing the bundle of r-jets, but perhaps it is simpler to define a neighborhood basis for Diff$^r(M)$ at f by means of charts. Choose a finite cover of M by charts. Every point x of M is contained in a ball of radius r_x, which is contained in one chart. The balls $B(x, r_x/2)$ cover M, and we may take a finite subcover $B(x_1, r_{x_1}/2), \ldots, B(x_n, r_{x_n}/2)$. Let $\delta = \inf(r_{x_i}/2)$. We can now be sure that if two functions f and g satisfy $d_0(f, g) < \delta$, then for every x in M there is a chart containing both $f(x)$ and $g(x)$ (see Figure 1.2). We can now calculate the C^r distance between f and g in a chart by calculating the maximum difference between corresponding derivatives of f and g expressed in the same chart.

Recall that for a complete metric space we have the Baire property: a countable intersection of open dense sets is dense. We say that a set is residual or generic if it contains a countable intersection of open dense sets; a property which holds on a residual set is said to be generic.

We can now reformulate two questions of utmost importance:

(1) Does $\overline{\text{Per}(f)} = \Omega(f)$ generically in Diff$^r(M)$?
(2) Does $R(f) = \Omega(f)$ generically in Diff$^r(M)$?

A difficult partial result in this direction is the Closing Lemma:

Theorem 1.8 (Pugh). $\overline{\text{Per}(f)} = \Omega(f)$ *generically in* Diff$^1(M)$.

We also know that generically, for homeomorphisms of a compact smooth manifold, $\Omega(f) = \overline{\text{Per}(f)} = R(f)$. On the other hand, the necessary perturbation theory is as yet unavailable in the case $r > 1$, because it is difficult to keep control of the higher derivatives.

Until now, by a homeomorphism (resp. diffeomorphism) of M we have meant a surjective mapping with a continuous (resp. differentiable) inverse, but this hypothesis of surjectivity is not really necessary. All our definitions still make sense when f is no longer surjective; it suffices simply not to consider $f^{-1}(x)$ for x not in the image of f. For example, $\bigcap_{n\geq 0} f^n(M)$ is a closed invariant set in M and

$$\mathrm{Per}(f) \subset L_+(f) \subset \Omega(f) \subset R(f) \subset \bigcap_{n\geq 0} f^n(M).$$

$L_-(f)$ is always the set of limit points of sequence $f^{-n}(x)$, where all the terms of the sequence are defined; $L_-(f) \subset \bigcap_{n\geq 0} f^n(M)$.

From now on, for the sake of simplicity, unless otherwise stated we will assume that M is compact, connected, and without boundary.

It is often useful to permit M to have a boundary, even though we will not check the corresponding versions of our theorems for this case. When M has a boundary we must assume that f is an embedding of M in its interior. For example, M might be a bounded region in a Euclidean space with smooth boundary. The reader who is unfamiliar with smooth manifolds may think of the case of bounded regions in Euclidean space.

Exercise 1.1. Show that if x belongs to Ω, and U is a neighborhood of x, there is a sequence of integers n_i tending to infinity such that the intersection $f^{n_i}(U) \cap U$ is nonempty.

Commentary

We think of a diffeomorphism $f: M \to M$ as a dynamical system, that is, as a differentiable action of the group $\mathbb{Z}$ on M. In general, for a Lie group G, a dynamical system of class $C^{s,r}$ is a group homeomorphism $\Phi: G \to \mathrm{Diff}^r(M)$ such that the natural map $G \times M \to M$, $(g, m) \mapsto \Phi(g)(m)$ has continuous partial derivatives of all orders up to s with respect to the variables of G, and up to r with respect to M; moreover, we require that the mixed partials of order up to $\rho = \min(r, s)$ be continuous as well. Such a map $G \times M \to M$ is C^ρ, each restriction to $G \times \{m\}$ is C^s, each restriction to $\{g\} \times M$ is C^r. Often r and s differ. In this case we speak of the action as being C^ρ. The fundamental examples come from the theory of ordinary differential equations on manifolds. Recall that an ordinary differential equation on a closed manifold (compact, no boundary) is given by a vector field $V: M \to TM$, that is a section of the tangent bundle. A solution of the equation passing through the point x_0 of M is an M-valued function $\varphi(x_0, t)$ defined for t in some open neighborhood of 0 in $\mathbb{R}$, satisfying $\varphi(x_0, 0) = x_0$ and $V(\varphi(x_0, u)) = (d\varphi(x_0, t)/dt)_{t=u}$. If V is C^r for some $r \geq 1$, then there is a unique solution $\varphi(x_0, t)$ defined for all

t in $\mathbb{R}$, which has the additional property that $\varphi(\varphi(x_0, t), s) = \varphi(x_0, t + s)$ for all t and s. Moreover, the map $\Phi: \mathbb{R} \to \text{Diff}^r(M)$, defined by $\Phi(t)(x) = \varphi(x, t)$ is a $C^{r+1,r}$ dynamical system, that is a $C^{r+1,r}$ action of $\mathbb{R}$ on M, which is also called a $C^{r+1,r}$ flow, or more usually, a C^r flow. [1.1] and [1.7] are two excellent elementary texts on ordinary differential equations. We attempt to make a qualitative study of the asymptotic behavior of the orbits of the actions and to link these asymptotic behaviors to stability. By asymptotic behavior of the orbit $\Phi(t)(x)$, we mean the behavior as t tends to $\pm\infty$ when the group is $\mathbb{Z}$ or $\mathbb{R}$. When G is an arbitrary Lie group, we mean by this the behavior of sequences $\Phi(g_i)(x)$, where (g_i) is a sequence which eventually leaves every compact subset of G. This approach is only meaningful when G is noncompact. The case where G is a compact Lie group is the object of study in compact differential geometry; in this context there is a strong stability theorem [1.18]. Asymptotic methods essentially permit us to understand the two cases of continuous time ($G = \mathbb{R}$) and discrete time ($G = \mathbb{Z}$). These two cases have numerous similarities. I will limit myself to the case where $G = \mathbb{Z}$, that is to the study of diffeomorphisms of M. The action of groups other than $\mathbb{R}$ or $\mathbb{Z}$ are studied in [1.3], [1.6], [1.13], and [1.14].

My point of view comes from [1.16]. In fact, the first four chapters recapitulate [1.16], the work through which I discovered the subject.

The concepts introduced in this first chapter all come from [1.16], with the exception of the concept of ε-pseudo-orbits which is found in [1.4] and [1.2] and that of the chain-recurrent set, first defined in [1.4] and developed in [1.5] and [1.19]. The generic point of view first appeared in the study of dynamical systems with the theorems of Peixoto on the openness and density of structurally stable flows on surfaces [1.11] and Kupka and Smale on the genericity of what are now called Kupka–Smale diffeomorphisms [1.8] and [1.17]. Along the same lines we have the proof of Pugh's Theorem in [1.12].

There is another way of relating flows and difeomorphisms–the suspension construction. Given a diffeomorphism $f: M \to M$, we define a map $\psi_s: M \times \mathbb{R} \to M \times \mathbb{R}$ by $\psi_s(m, t) = (m, t + s)$; ψ is a flow on $M \times \mathbb{R}$. The quotient of $M \times \mathbb{R}$ by the relations $(m, t) \sim (f(m), t + 1)$ is compact. This quotient space $\tilde{M} = M \times \mathbb{R}/\sim$ is a bundle over S^1 with fiber diffeomorphic to M. The flow ψ_s induces a flow Ψ_s on $\tilde{M}$. The dynamics of Ψ_s and f are intimately linked.

Some good references on dynamical systems are [1.2], [1.9], and [1.10]. I have borrowed broadly from these works.

References

[1.1] Arnold, V. I., *Ordinary Differential Equations*, MIT Press, Cambridge, Mass., 1978.

[1.2] Bowen, R., *Equilibrium States and the Ergodic Theory of Axiom A Diffeomorphisms*, Lecture Notes in Mathematics, No. 470, Springer-Verlag, New York, 1975.

[1.3] Camacho, C., On $R^k \times Z^1$-actions, in *Dynamical Systems*, Peixoto (Ed.), Academic Press, New York, 1973, p. 23.
[1.4] Conley, C., On the gradient structure of a flow, IBM Preprint.
[1.5] Conley, C., Some aspects of the qualitative theory of differential equations, in *Dynamical Systems*, Proceedings of an International Symposium, Cesari, Hale and Lasalle (Ed.), Academic Press, New York, 1976.
[1.6] Hirsch, M. W., Foliations and non-compact transformation groups, *Bull. Amer. Math. Soc.* **76** (1970), 1020.
[1.7] Hirsch, M. W. and Smale, S., *Differential Equations, Dynamical Systems and Linear Algebra*, Academic Press, New York, 1974.
[1.8] Kupka, I., Contributions à la théorie des champs générique, *Contrib. Differential Equations*, **2** (1963), 437.
[1.9] Melo, W. and Palis, J., *Geometric Theory of Dynamical Systems*, Springer-Verlag, New York, 1982.
[1.10] Nitecki, Z., *Differentiable Dynamics*, MIT Press, Cambridge, Mass., 1972.
[1.11] Peixoto, M., Structural stability on two-dimensional manifolds, *Topology* **1** (1962), 101
[1.12] Pugh, C. C., An improved closing lemma and a general density theorem, *Amer. J. Math.* **89** (1967), 1010.
[1.13] Pugh, C. C. and Shub, M., Ergodicity of Anosov actions, *Invent. Math.* **15** (1972), 1.
[1.14] Pugh, C. C. and Shub, M., Axiom A actions, *Invent. Math.* **29** (1974), 7.
[1.15] Shub, M., Dynamical systems, filtrations, and entropy, Bull. Amer. Math. Soc. **80** (1974), 27.
[1.16] Smale, S., Differentiable dynamical systems, *Bull. Amer. Math. Soc.* **73** (1967), 747.
[1.17] Smale, S., Stable manifolds for differential equations and diffeomorphisms, *Ann. Scuola Norm. Sup. Pisa* **17** (1963), 97.
[1.18] Palais, R., Equivalence of nearby differential actions of a compact group, *Bull. Amer. Math. Soc.* **67** (1961), 362.
[1.19] Bowen, R., *On Axiom A Diffeomorphisms*, American Mathematical Society, Providence, R.I., 1978.
[1.20] Smale, S., *The Mathematics of Time*, Springer-Verlag, New York, 1980.
[1.21] Franks, J., *Homology and Dynamical Systems*, American Mathematical Society, Providence, R.I., 1972.
[1.22] Irwin, M. C., *Smooth Dynamical Systems*, Academic Press, New York, 1980.
[1.23] Guckenheimer, J. and Holmes, P., *Nonlinear Oscillations, Dynamical Systems and Difurcation of Vector Fields*, Springer-Verlag, New York, 1983.

CHAPTER 2
Filtrations

One way to understand the recurrence of a homeomorphism is to try to build up the underlying manifold from simpler pieces, each of which isolates and asymptotically specifies an invariant set. With this in mind we make several definitions. Given a homeomorphism f, of M, a filtration $\mathbf{M}$ adapted to f is a nested sequence $\varnothing = M_0 \subset M_1 \subset \cdots \subset M_k = M$ of smooth, compact codimension 0 submanifolds with boundary of M, such that $f(M_i), \subset \operatorname{Int} M_i$. We denote by $\mathbf{M}^{-1}$ the filtration

$$\varnothing = \overline{M - M_k} \subset \overline{M - M_{k-1}} \subset \cdots \subset \overline{M - M_0} = M$$

which is adapted to f^{-1}.

For a filtration $\mathbf{M}$ adapted to f,

$$K_\alpha^f(\mathbf{M}) = \bigcap_{n \in \mathbb{Z}} f^n(M_\alpha - M_{\alpha-1})$$

is compact and is the maximal f-invariant subset of $M_\alpha - M_{\alpha-1}$. When no confusion looms, we will often suppress reference to f. Now, for the sake of comparison, we let $K^f(\mathbf{M}) = \bigcup_{\alpha=1}^{k} K_\alpha^f(\mathbf{M})$ and set

$$\Omega_\alpha = \Omega \cap M_\alpha - M_{\alpha-1}, \qquad R_\alpha = R \cap M_\alpha - M_{\alpha-1}.$$

Ω_α and R_α are contained in $K_\alpha(\mathbf{M})$, so $\Omega \subset K(\mathbf{M})$ and, consequently, $L \subset K(\mathbf{M})$.

Now we seek to determine which closed invariant sets are of the form $K(\mathbf{M})$, for some filtration $\mathbf{M}$ adapted to f.

Definition 2.1. For a subset B of M, *the stable and unstable sets* W^s and W^u of B are

$$W^s(B) = \{y \in M \mid d(f^n(y), f^n(B)) \to 0 \text{ as } n \to +\infty\},$$

$$W^u(B) = \{y \in M \mid d(f^n(y), f^n(B)) \to 0 \text{ as } n \to -\infty\}.$$

Consider a disjoint union $\Lambda = \Lambda_1 \cup \cdots \cup \Lambda_r$ of closed f-invariant sets, such that $L(f) \subset \Lambda$.

Lemma 2.2. *M is a disjoint union of the $W^s(\Lambda_i)$'s and also of the $W^u(\Lambda_i)$'s.*

PROOF. Choose open neighborhoods $\mathbf{U}_i$ of Λ_i, such that $f(\mathbf{U}_i) \cap \mathbf{U}_j = \varnothing$ for $i \neq j$. For any x in M, $f^k(x)$ is contained in $\bigcup_{i=1}^r \mathbf{U}_i$, for large k, since the sequence $f^n(x)$ cannot accumulate in the complement $M - \bigcup_{i=1}^r \mathbf{U}_i$; which is compact. Moreover, the sequence $f^n(x)$ is eventually contained in one particular $\mathbf{U}_k$, since $f(\mathbf{U}_i) \cap \mathbf{U}_j = \varnothing$, for $i \neq j$. Choosing $\mathbf{U}_i$'s arbitrarily close to the Λ_i's we see $x \in W^s(\Lambda_k)$ for some k. □

We may define a preorder $\gg$, on the Λ_i's by

$$\Lambda_i \gg \Lambda_j \quad \text{iff} \quad (W^u(\Lambda_i) - \Lambda_i) \cap (W^s(\Lambda_j) - \Lambda_j) \neq \varnothing.$$

We say the preorder has an r-cycle if there is a sequence $\Lambda_{i_1} \gg \cdots \gg \Lambda_{i_{r+1}} = \Lambda_{i_1}$. If $\gg$ has no cycles, it may be extended to a total ordering $>$; that is, for all i and j either $\Lambda_i < \Lambda_j$, $\Lambda_i = \Lambda_j$, or $\Lambda_i > \Lambda_j$. Where the first inequality holds there is no sequence $\Lambda_i \gg \cdots \gg \Lambda_j$. Note that in this case we may reindex so that $\Lambda_i > \Lambda_j$ iff $i > j$; we call such an ordering of the Λ_i's a *filtration ordering.*

Theorem 2.3. *Given a homeomorphism f of M, and disjoint union $\Lambda = \Lambda_1 \cup \cdots \cup \Lambda_r$ of closed invariant sets with $L(f) \subset \Lambda$; in order that $\Lambda_i = K_i(\mathbf{M})$ for $i = i, \ldots, r$, for some filtration $\mathbf{M}$ adapted to f, it is necessary and sufficient that the Λ_i's have no cycles and the ordering by indices be a filtration ordering.*

PROOF. We first demonstrate necessity. Observe that if M is adapted to f, then $f(M_i)$ is contained in the interior of M_i, so $W^u(K_i(\mathbf{M}))$ is contained in M_i, and $W^s(K_i(\mathbf{M}))$ is contained in $M - M_{i-1}$. Thus we cannot have $K_j(\mathbf{M}) \gg \cdots \gg K_i(\mathbf{M})$ if $j < i$. Since $K_i(\mathbf{M}) = \bigcap_{n \in \mathbb{Z}} f^n(M_i - M_{i-1})$, the observation above also shows, since W^u and W^s are evidently f invariant, that $W^u(K_i(\mathbf{M})) \cap W^s(K_i(\mathbf{M})) = K_i(\mathbf{M})$. Thus we cannot have $K_i(\mathbf{M}) \gg K_i(\mathbf{M})$ either. □

To demonstrate sufficiency we need a series of lemmas.

Lemma 2.4. *If $\overline{W^u(\Lambda_i)}$ intersects $W^u(\Lambda_j)$ then it intersects Λ_j.*

PROOF. Since $\overline{W^u(\Lambda_i)}$ is a closed f-invariant set, all the limit points of $f^n(y)$ for y in $W^u(\Lambda_j) \cap \overline{W^u(\Lambda_i)}$ must be in $\overline{W^u(\Lambda_i)}$, but the α limit points of an orbit in $W^u(\Lambda_j)$ must actually be in Λ_j. By compactness such α limit points of $\{f^n(y)\}$ exist so $\overline{W^u(\Lambda_i)} \cap \Lambda_j \neq \varnothing$. □

Lemma 2.5. *If $\overline{W^u(\Lambda_i)}$ intersects Λ_j, for $i \neq j$, then it intersects $W^s(\Lambda_j) - \Lambda_j$.*

PROOF. Choose small compact sets $\mathbf{U}_k$ which contain the Λ_k's in their interior with the property that $f(\mathbf{U}_k) \cap \mathbf{U}_m = \varnothing$ for $k \neq m$. Now choose a sequence of points of distinct orbits x_n in $W^u(\Lambda_i)$ converging to Λ_j; by compactness of $\mathbf{U}_j$ it is no loss of generality to choose them in $\text{Int}(\mathbf{U}_j)$. Now the definition of $W^u(\Lambda_i)$ allows us to find, for each x_n, a least positive integer k_n, with the property that $f^{-k_n}(x_n)$ is not in the interior of $\mathbf{U}_j$. Let x be a limit point of the sequence $f^{-k_n}(x_n)$ and notice $x \in W^u(\Lambda_i)$. Since $x_n \to \Lambda_j$ as $n \to \infty$, $k_n \to \infty$. Finally, we claim that $f^m(x)$ is in $\text{Int}(\mathbf{U}_j)$ for all positive m; since $\mathbf{U}_j$ is small, yet x is not in $\text{Int}(\mathbf{U}_j)$ we see x is also in $W^s(\Lambda_j) - \Lambda_j$ and the conclusion follows. If there were a positive m with $f^m(x)$ in $M - \text{Int}(\mathbf{U}_j)$, then there would be an x_n near x with $k_n > m$ and we would have $f^{-(k_n-m)}(x_n)$ in $M - \mathbf{U}_j$ contrary to the minimality of k_n. □

Lemma 2.6. *If $\overline{W^u(\Lambda_i)}$ intersects Λ_j, for $i \neq j$, then i is greater than j.*

Proof. By Lemma 2.5 we can find an x in $\overline{W^u(\Lambda_i)} \cap (W^s(\Lambda_j) - \Lambda_j)$ and from Lemma 2.2. we can find an m_1, with $\Lambda_{m_1} > \Lambda_j$ and $x \in W^u(\Lambda_{m_1})$. Therefore, by Lemma 2.4, $\overline{W^u(\Lambda_i)} \cap \Lambda_{m_1}$ is nonempty. Repeating this reasoning we construct a sequence $\cdots > \Lambda_{m_r} > \cdots > \Lambda_{m_1} > \Lambda_j$, where $\overline{W^u(\Lambda_i)} \cap \Lambda_{m_r}$ is nonempty. Since there are no cycles and only a finite number of distinct Λ_k's we eventually find an r such that $\Lambda_{m_r} = \Lambda_i$. □

Lemma 2.7. $\overline{W^u(\Lambda_i)} \subset \bigcup_{j \le i} W^u(\Lambda_j)$, *hence* $\bigcup_{j \le i} W^u(\Lambda_j)$ *is closed.*

Proof. Let x be in $\overline{W^u(\Lambda_i)}$. By Lemma 2.2, x is in $W^u(\Lambda_j)$, for some j, and Lemmas 2.4 and 2.6 guarantee $i \ge j$. □

Lemma 2.8. $\bigcup_{j \le i} W^s(\Lambda_j)$ *is an open neighborhood of* $\bigcup_{j \le i} W^u(\Lambda_j)$.

Proof. Applying Lemma 2.7 to f^{-1} and remembering that this interchanges W^u and W^s and reverses the order, we see $\bigcup_{j \le i} W^s(\Lambda_j)$ is open and contains $\bigcup_{j \le i} W^u(\Lambda_j)$. □

Lemma 2.9. *Let X be a compact metric space and $f: X \to X$ a homeomorphism onto its image. Consider a compact invariant set P of the form $P = \bigcap_{n \ge 0} f^n(Q)$, where Q is a compact set containing P in its interior. If f is open on a neighborhood of P then there exists a compact neighborhood $\mathbf{U}$ of P, contained in Q, which is mapped into its interior by f.*

Proof. Set $A_r = \bigcap_{n=0}^{r} f^n(Q)$. The A_r's are a nested sequence of compact sets converging to P; in particular, $A_r \subset \text{Int}(Q)$ for large r. Choose an intermediate set $\mathbf{U}$ where f is open such that $P \subset \text{Int}(\mathbf{U}) \subset \mathbf{U} \subset Q$ and also $f(\mathbf{U}) \subset Q$, for instance $\mathbf{U} = Q \cap f^{-1}(Q)$. Fix an r large enough that A_r is in the interior of $\mathbf{U}$ and note that $f(A_r) \subset f(\mathbf{U}) \subset Q$ so $f(A_r) \subset A_r$. Since A_r is a neighborhood of P, there is an n so large that $f^n(A_r)$ is contained in the interior of A_r. If $n = 1$, we are done; if not we can judiciously choose a slightly larger compact neighborhood $\hat{E}$ which is mapped into its interior. To see this we assume $n \ge 2$

and construct an E mapped into its interior by f^{n-1}; $\hat{E}$ is produced by descent. Choose a small compact neighborhood W of A_r with $A_r \subset \text{Int } W \subset U$ and $f^n(W) \subset \text{Int } A_r$. Let $E = f^{n-1}(W) \cup A_r$; $f^{n-1}(A_r) \subset \text{Int}(f^{n-1}(W)) \subset \text{Int } E$ and $f^{n-1}(f^{n-1}(W)) \subset f^{n-2}(f^n(W)) \subset f^{n-2}(\text{Int } A_r) \subset \text{Int } A_r \subset \text{Int } E$, so E is as claimed. □

We now finish the proof of the theorem by constructing the desired filtration. First notice that if K is a compact subset of $\bigcup_{j=1}^{P} W^s(\Lambda_{i_j})$ then

$$\bigcap_{n\geq 0} f^n(K) \subset \bigcup_{j=1}^{P} W^u(\Lambda_{i_j}).$$

For if $x \in \bigcap_{n\geq 0} f^n(K)$, then $f^{-n}(x) \in K$, for all positive n so $L_-(x) \subset K$, and consequently $x \in W^u(\Lambda_k)$ for some k. Thus Λ_k intersects $\bigcup_{j=1}^{P} W^s(\Lambda_{i_j})$ and so must be one of the Λ_{i_j}'s.

We will construct M_1 and continue the construction stepwise as follows:

(1) Λ_1 is at the bottom of the order so $W^s(\Lambda_1)$ is a neighborhood of Λ_1. Choose a compact neighborhood Q of Λ_1 contained in $W^s(\Lambda_1)$. We then have $\Lambda_1 = \bigcap_{n\geq 0} f^n(Q)$, so, by Lemma 2.9, there's a compact neighborhood V_1 of Λ_1 mapped into its interior. Now choose a smooth, non-negative function g such that $V_1 = \{x \mid g(x) = 0\}$; for small enough ε, $M_1 = g^{-1}[0, \varepsilon]$ is a smooth manifold with boundary with $f(M_1) \subset \text{Int}(M_1)$.

(2) Lemma 2.8 tells us that $W^s(\Lambda_2) \cup W^s(\Lambda_1)$ is an open neighborhood of $W^u(\Lambda_2) \cup W^u(\Lambda_1)$. Then $W^u(\Lambda_2) \cup W^s(\Lambda_1) = \bigcap_{n\geq 0} f^n(Q_2)$, so Lemma 2.9 again allows us to find a compact neighborhood V_2 mapped into its interior and, approximating $V_2 \cup M_1$ as in (1), to produce a submanifold M_2 containing $V_2 \cup M_1$ which is also mapped into its interior. Proceeding inductively we produce a filtration M, adapted to f.

(3) Finally, we check that $\Lambda_j = \bigcap_{n\in\mathbb{Z}} f^n(M_j - M_{j-1})$. By construction we have

$$\bigcap_{n\geq 0} f^n(M_j) = \bigcup_{i\leq j} W^u(\Lambda_i)$$

and

$$\bigcup_{i<j} W^u(\Lambda_i) \subset M_{j-1}$$

so

$$\bigcap_{n\geq 0} f^n(M_j - M_{j-1}) \subset W^u(\Lambda_j).$$

On the other hand, by considering $\mathbf{M}^{-1}$, we have $\bigcap_{n\leq 0} f^n(M_j - M_{j-1}) \subset W^s(\Lambda_j)$. $W^u(\Lambda_j) \cap W^s(\Lambda_j) = \Lambda_j$ so we are done. □

Filtrations can tell us a lot about stability; they are robust and indicate the persistence of certain invariant sets.

Proposition 2.10. *Let* $\mathbf{M}$ *be a filtration adapted to* f *and* $\mathbf{U}$ *be a neighborhood of* $K^f(\mathbf{M})$. *There is a neighborhood* V *of* f *in* $\mathrm{Hom}(M)$ *such that* $\mathbf{M}$ *is adapted to all* g *in* V *and* $K^g(\mathbf{M})$ *is in* $\mathbf{U}$. *Furthermore,* V *may be chosen so that* $K^g_\alpha(\mathbf{M})$ *is contained in* $\mathbf{U}_\alpha = (M_\alpha - M_{\alpha-1}) \cap \mathbf{U}$.

Proof. Since $f(M_\alpha)$ is contained in the interior of M_α the same holds for any C^0 close map g. Similarly, since $\mathbf{U}_\alpha$ is a neighborhood of $K^f_\alpha(M)$ so there is an N such that $\bigcap_{-N}^{N} f^i(M_\alpha - \mathrm{Int}\, M_{\alpha-1}) \subset \mathbf{U}_\alpha$, we also have $\bigcap_{-N}^{N} g^i(M_\alpha - \mathrm{Int}\, M_{\alpha-1}) \subset \mathbf{U}_\alpha$ for g C^0 close to f. □

Commentary

Smale [1.16] uses filtrations to control the growth of the nonwandering set under perturbations. The order conditions were used by Smale [2.7] when he established the Morse inequalities for what are now known as Morse–Smale diffeomorphisms. Rosenberg [2.4] has generalized these conditions to define the no-cycle property and extend the Morse inequalities. Palis [2.3] defined filtrations which allowed him to show that the Morse–Smale diffeomorphisms of a manifold are open, and I owe to him the sequence of lemmas used to prove Theorem 2.3. He described them to me when we were both working under the direction of Smale at Berkeley between 1964 and 1967. Theorem 2.3 and Proposition 2.10 come essentially from [1.16] and have been restated in more recent works.

Newhouse [2.1] has studied filtrations in a more general context than Smale, while purusing his study of limit sets and filtration orderings.

These ideas are systematized in [2.6], [2.5], and [2.2].

Conley, [1.4] and [1.5], has considered sequences of filtrations in the topological case and had the idea to apply them to the study of the chain-recurrent set.

References

[2.1] Newhouse, S., On hyperbolic limit sets, *Trans. Amer. Math. Soc.* **167** (1972), 125.

[2.2] Nitecki, Z. and Shub, M., Filtrations, decompositions, and explosions, *Amer. J. Math.* **107** (1975), 1029.

[2.3] Palis, J., On Morse–Smale dynamical systems, *Topology* **8** (1969), 385.

[2.4] Rosenberg, H., A generalization of the Morse–Smale inequalities, *Bull Amer. Math. Soc.* **70** (1964), 422.

[2.5] Shub, M., Stability and genericity for diffeomorphisms, in *Dynamical Systems*, Peixoto (Ed.), Academic Press, New York, 1973, 493.

[2.6] Shub, M. and Smale, S., Beyond hyperbolicity, *Ann. of Math.* **96** (1972), 587.

[2.7] Smale, S., Morse inequalities for a dynamical system, *Bull. Amer. Math. Soc.* **66** (1960), 43.

CHAPTER 3

Sequences of Filtrations

Definition 3.0. Let $\mathbf{M}$: $\varnothing = M_0 \subset \cdots \subset M_k = M$ and $\mathbf{N}$: $\varnothing = N_0 \subset \cdots \subset N_l = M$ be two filtrations of M. We say that N *refines* M if and only if for all $i = 0, \ldots, l-1$, there is a j, $0 \leq j \leq k-1$ such that $(N_{i+1} - N_i) \subset (M_{j+1} - M_j)$.

If $\mathbf{M}$ and $\mathbf{N}$ are two filtrations adapted to the same homeomorphism f of M and $\mathbf{N}$ refines $\mathbf{M}$, $K^f(\mathbf{N})$ is clearly contained in $K^f(\mathbf{M})$. This is the utility of sequences of filtrations; even when it is impossible, given a closed invariant set Λ, to find a filtration $\mathbf{M}$ with $K^f(\mathbf{M}) = \Lambda$, one can often find a sequence of $\mathbf{M}_i$'s with $K^f(\mathbf{M}_i)$ converging to Λ.

Definition 3.1. $\mathbf{M}_1, \mathbf{M}_2, \ldots$ is a *sequence of filtrations* if $\mathbf{M}_i$ refines $\mathbf{M}_{i-1}$ for all i. Let $K\{\mathbf{M}_i\} = \bigcap_i K(\mathbf{M}_i)$. If Λ is a closed invariant set for f and if $\mathbf{M}$ is a filtration adapted to f such that $K(M) = \Lambda$ we say that $\mathbf{M}$ is a *filtration* for Λ. If $\{\mathbf{M}_i\}$ is a sequence of filtrations adapted to f such that $K\{\mathbf{M}_i\} = \Lambda$, we say that $\{\mathbf{M}_i\}$ is a sequence of filtrations for Λ. A filtration for Ω is called a *fine filtration*. A sequence of filtrations for Ω is called a *fine sequence of filtrations*.

We will prove that there is always a sequence of filtrations for $R(f)$, after first giving some supplemental technical properties of filtrations.

The following refines Theorem 2.3.

Proposition 3.2. *Suppose that $\Lambda = \Lambda_1 \cup \cdots \cup \Lambda_k$ is a disjoint union of closed invariant sets for f which contain $L(f)$ and, moreover, that the Λ_i's have no cycles. Let* $\mathbf{N}$ *be a filtration adapted to f such that, for every i, there is an α such that Λ_i is contained in $K^f_\alpha(\mathbf{N})$. Then there is a filtration for Λ which refines* $\mathbf{N}$.

PROOF. As in the proof of Theorem 2.3, if Λ_i is contained in $\mathbf{N}_\alpha - \mathbf{N}_{\alpha-1}$, we can construct a new $\mathbf{M}_i$ containing $\mathbf{N}_{\alpha-1}$ and contained in $\mathbf{N}_\alpha$. We can thus insert the $\mathbf{N}_\alpha$'s in the sequence of $\mathbf{M}_i$'s and the resulting filtration will clearly refine $\mathbf{N}$. □

Let Λ be a closed invariant set for a homeomorphism f, and let $W_1, \ldots, W_K$ be pairwise disjoint open sets covering Λ: $\Lambda \subset \bigcup_{i=1}^K W_i$. Now we can define $B(W_i) = \bigcup_{n\geq 0} f^{-n}(W_i)$ and $F(W_i) = \bigcup_{n\geq 0} f^n(W_i)$, the backwards and forwards of W_i. Let $K_i = \bigcap_{n\in\mathbb{Z}} f^n(W_i)$ be the maximal f invariant set in W_i, and define a preorder $\gg$ in W_i by

$$W_i \gg W_j \begin{cases} i \neq j & \text{and} \quad F(W_i) \cap B(W_j) \neq \emptyset, \\ i = j & \text{and} \quad B(W_i) - W_i \cap F(W_i) - W_i \neq \emptyset. \end{cases}$$

The set of W_i's contains an r-cycle if there is a sequence $W_{i_1} \gg \cdots \gg W_{i_{r+1}} = W_{i_1}$. As usual, if the W_i's have no cycles, we say that the order of indices is a filtration ordering if whenever j is greater than i there is no sequence $W_i \gg \cdots \gg W_j$.

Proposition 3.3. *Let Λ be a closed invariant set for a homeomorphism f of M, and assume that Λ contains $L(f)$. Let $W_1, \ldots, W_k$ be a pairwise disjoint open cover of Λ which has no cycles and whose indices induce a filtration ordering. Then the sets K_i defined above are disjoint closed invariant sets for f and have no cycles. Moreover, $L(f)$ is contained in the union of the K_i's.*

PROOF. First we show that the K_i's are closed. Let x be a point in the closure of K_i. Since $\omega(x) \subset L(f) \subset \Lambda \subset \bigcup_{i=1}^k W_i$, there is an integer N so large that for n greater than or equal to N, $f^n(x)$ belongs to $\bigcup_{i=1}^k W_i$. Furthermore, since the W_i's have no cycles we can find j and N' such that for all $n \geq N'$, $f^n(x)$ belongs to a particular W_j. We must also have $j = i$; since W_j and $f^{-n}(W_j)$ are open, if $f^{-n}(W_j)$ contains a point of $\bar{K}_i$, it must also contain points in K_i.

Now check the inclusion of $L(f)$ in the union of the K_i's. Since there are no 1 cycles, when x is in K_i so is $f^n(x)$, for all n. For arbitrary x in M, since $\omega(x)$ is in Λ, $f^n(x)$ must be in one of the W_i's, W_j, say, for sufficiently large n. Thus the ω-limit points of $f^n(x)$ and hence of x must be in K_j. □

Theorem 3.4. *For any homeomorphism f there are sequences of filtrations for $R(f)$.*

PROOF. Let $\varepsilon > 0$ and take $0 < \alpha < \varepsilon/2$ so small that $d(x, y) < \alpha$ implies $d(f(x), f(y)) < \varepsilon/2$. Cover Ω by open balls of radius $\alpha/2$, centered at points of Ω and extract a finite subcover $\{B_i = B(x_i, \alpha/2), i = 1, \ldots, k\}$. Define a preorder on the B_i's by $B_i \gg B_j$ whenever there is a nonnegative n such that $f^n(B_i) \cap B_j$ is nonempty. We say that two balls are equivalent if they belong to the same cycle for this preorder. Now regroup the equivalence classes of B_i's into open sets $U_j (j = 1, \ldots, q)$ and let $V_j = F(U_j) \cap B(U_j)$. The V_j's form a pairwise disjoint open cover of Ω with no cycles.

Notice that all the points in one V_i belong to the same ε-pseudo-orbit. This is because if x is in B_i and y is in an equivalent B_j, we can find an n such that $f^n(B_i) \cap B_j$ is nonempty and we have two simple cases:

(1) $n > 0$. Choose a z in B_i with $f^n(z)$ also in B_j; then $\{x, f(z), \ldots, f^{n-1}(z), y\}$ is an ε-pseudo-orbit by our choice of α.
(2) $n = 0$. Since x_i is in Ω we can find an $n \geq 2$ such that $f^n(B_i)$ intersects B_i. Choose z in B_i with $f^n(z)$ in B_i. The sequence $\{x, f(z), \ldots, f^{n-1}(z), y\}$ is the ε-pseudo-orbit we seek.

From this it also follows that any point in V_i is ε-pseudoperiodic.

Now if we define sets $K_i = \bigcap_{n=\mathbb{Z}} f^n(V_i)$ as above, every point in a K_i is also ε-pseudoperiodic. There is an obvious filtration for the $K = \bigcup_{i=1}^{q} K_i$ which implies, letting ε tend to 0, that $R(f)$ is contained in K. After wiggling the V_i's a little we may suppose their closures are disjoint and that $\bigcup_i V_i$ is a neighborhood of Ω. Now let ε_n be a sequence of positive numbers tending to 0 and for each ε_n construct V_i^n, K_i^n, and K^n as above. Taking care to choose the sequence ε_n so that the neighborhood $(\bigcup_i V_i^{n+1})$ of Ω is contained in $(\bigcup_i V_i^n)$, for all n, we see that each K_i^{n+1} is contained in some K_j^n, since each V_i^{n+1} is contained in a V_j^n (otherwise there is a cycle of ε_{n+1} balls not contained in a V_i^n and hence a cycle in the V_i^n's).

Proposition 3.2 allows us to refine the filtration for the K_i^n's to one for the K_i^{n+1}'s. The K^n's thus consist of ε_n-pseudoperiodic points and each contains $R(f)$; therefore, $\bigcap_{n \geq 0} K^n = R(f)$. □

Consider a filtration $M = M_2 \supset M_1 \supset M_0 = \varnothing$ adapted to a homeomorphism f. We can easily construct a continuous nonnegative function φ such that

$$\varphi = 1 \text{ on } K_2; \qquad \varphi = 0 \text{ on } K_1,$$

$$\varphi(f(x)) < \varphi(x) \qquad \text{for} \quad x \notin K.$$

To do this we define φ on $\bigcup_{n \in \mathbb{Z}} f^n(\partial M_1)$ by

$$\varphi(x) = \tfrac{1}{2} - \operatorname{sign}(n) \sum_{i=1}^{|n|} 2^{-(i+1)} \quad \text{if} \quad x \in f^n(\partial M).$$

The Tietze extension theorem then allows us to extend to a positive continuous function with level sets as shown in Figure 3.1.

Now consider a filtration with more than two elements, for example three: $M = M_3 \supset M_2 \supset M_1 \supset M_0 = \varnothing$. Then $M = M_3 \supset M_2 \supset M_0 = \varnothing$ and $M = M_3 \supset M_1 \supset M_0 = \varnothing$ are filtrations to which the above construction applies, and, by adding the functions it yields, we can find a function that is 2 on K_2, 1 on K_1, 0 on K_0, and decreases along orbits off K. This clearly works for any finite filtration, and by properly scaling and taking limits we get a continuous function decreasing along orbits off $K\{\mathbf{M}_i\}$ for any sequence of filtrations $\{\mathbf{M}_i\}$ adapted to f. In fact, with more work than we will do here, one can find smooth functions φ as above.

Together, this construction gives us:

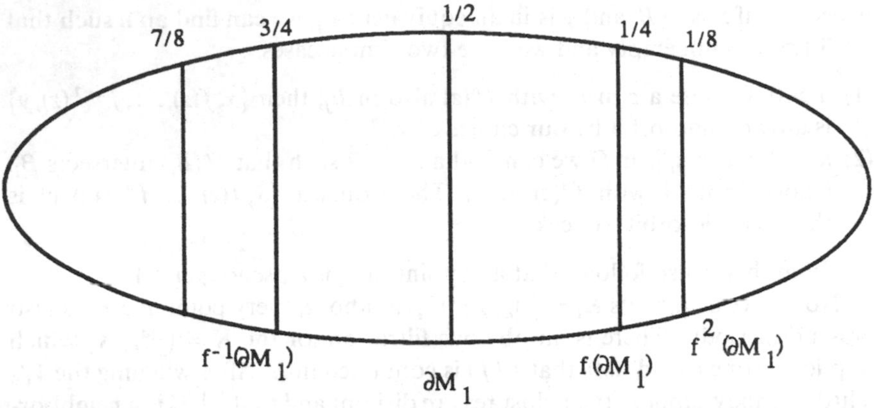

Figure 3.1.

Proposition 3.5. *Let* **M** *be a filtration (resp.* $\{\mathbf{M}_i\}$ *a sequence of filtrations) adapted to* f. *Then there is a continuous function* $\varphi\colon M \to \mathbb{R}$, *decreasing along orbits of* f *not in* $K(\mathbf{M})$ *(resp.* $K\{\mathbf{M}_i\}$*)*,

$$x \notin K(\mathbf{M})(\text{resp. } K\{\mathbf{M}_i\}) \Rightarrow \varphi(f(x)) < \varphi(x).$$

In particular, there is a function φ decreasing along orbits of f not in $R(f)$.

Proposition 3.6. *For any homeomorphism* f *of* M *we have* $R(f|_{R(f)}) = R(f)$.

This is not as ridiculous as it seems. The restriction of f to $R(f)$ sends $R(f)$ into $R(f)$, but it is not obvious that all pseudo-orbits starting in $R(f)$ can be made to stay in $R(f)$. Moreover, we can have $\Omega(f|_{\Omega(f)}) \neq \Omega(f)$, as in Figure 3.2.

Proof. The existence of a continuous φ decreasing along orbits off of R shows that for small α, α-pseudo-orbits starting in R do not stray too far from R. In order to find an ε-pseudoperiodic orbit contained in R, we can first find an α so small that any α-pseudoperiodic orbit starting from R lies in an $\varepsilon/2$ neighborhood of R. We then simply push it into R; the details are left to the reader. □

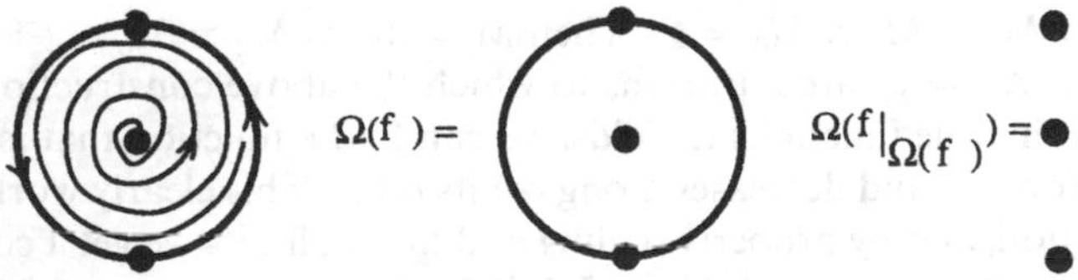

Figure 3.2. Here $\Omega(f)$ is the center and the outer circle and $\Omega(f|_{\Omega(f)})$ are the three marked points.

Proposition 3.7. *If U is a neighborhood of $R(f)$ there is a neighborhood V of f in* Hom(M) *such that* R(g) *is contained in U for all g in V.*

PROOF. Take a sequence of filtrations for $R(f)$. For large enough j, $K(M_j)$ is contained in U and we can use Proposition 2.4 to find V. □

Proposition 3.8. *A homeomorphism f of M has a fine sequence of filtrations if and only if $\Omega(f) = R(f)$.*

PROOF. If $\Omega(f) = R(f)$, we are done by Theorem 3.4. On the other hand, if Ω has a sequence of filtrations $\{M_i\}$, we have $R(f) \subset K\{M_i\} = \Omega(f)$, and we know $\Omega(f)$ is always contained in $R(f)$. □

Definition 3.9. We say a homeomorphism f of M *has no C^0 Ω explosions* if given a neighborhood U of $\Omega(f)$ we can find a neighborhood V of g with $\Omega(g) \subset U$ for all g in V.

Proposition 3.10. *If f has a fine sequence of filtrations, f does not have any C^0 Ω explosions.*

PROOF. This follows immediately from Propositions 3.7 and 3.8. □

The converse is also true:

Theorem 3.11. *For a homeomorphism f of M the following are equivalent:*

(1) *f has a fine sequence of filtrations;*
(2) *f does not have any C^0 Ω explosions;*
(3) $\Omega(f) = R(f)$.

All that is left to show is that (2) implies (3). We give a proof for manifolds of dimension 2 or greater. It depends on the following result from differential topology:

Proposition 3.12. *Let $\varepsilon > 0$ and let M be a smooth manifold of dimension greater than or equal to 2. Let (x_i, y_i) be pairs of points in M with $d(x_i, y_i) > \varepsilon$. Suppose that the x_i's are pairwise distinct, as are the y_i's. Then there is a diffeomorphism g of M taking x_i to y_i, which is $7\varepsilon - C^0$ close to the identity.* □

PROOF OF THEOREM FOR dim $M \geq 2$. If $\Omega(f)$ is different from $R(f)$, there is a neighborhood U of $\Omega(f)$ which does not contain all of $R(f)$. Choose $\varepsilon > 0$ an x in $R(f)$ outside of U. Then x is $\varepsilon/7$-pseudoperiodic so there is a sequence $\underline{x} = \{x = x_0, x_1, \ldots, x_k = x_0\}$ of length k such that $d(f(x_i), x_{i+1}) < \varepsilon/7$. We may take the x_i, and hence their images, to be distinct.

Let g be the diffeomorphism of Proposition 3.12 with $g(f(x_i)) = x_{i+1}$; $0 \leq i \leq k - 1$; and $d(g, \text{id}) < \varepsilon$. The point x is now an honest periodic point of $g \circ f$. Thus x is in $\Omega(g \circ f)$ and, since ε was arbitrary, also in $\Omega(f)$. □

The vector field case is similar and the results simpler to formulate. We will quickly give some of the corresponding definitions and theorems.

Definition 3.13. If X is a vector field on M and $\Lambda \subset M$ is a set invariant under the flow Φ induced by X, a smooth function $L: M \to R$ is a *Lyapunov function for* (X, Λ) if Λ is the critical set of L and the derivative $X(L)$ of L along X is negative on $M - \Lambda$.

Let δ and T be positive; we say that a point m of M is *(δ, T)-recurrent* or that the point m belongs to $R_{\delta,T}$ if there exist points y_i in M and numbers s_j, $0 \le j \le k$ such that

$$s_j > T; \qquad d(x_j, y_j) < \delta; \qquad x_0 = m; \qquad x_{j+1} = \Phi_{s_j}(y_j); \qquad y_k = m.$$

The *chain-recurrent set of* X is

$$R_X = \bigcap_{\delta, T} R_{\delta,T}.$$

Lyapunov functions for R_X correspond to fine sequences of filtrations for $R(f)$.

Theorem 3.14. *If X is a C^0 vector field with unique integral curves, there is a C^∞ Lyapunov function for (X, R_X).* □

Given a manifold N and a map $\varphi: N \to x^r(M)$, the C^r vector fields on M, one hopes to find a map $\alpha: N \to C^\infty(M)$ such that for every $n \in N$, $\alpha(n)$ is a Lyapunov function for $(\varphi(n), R\varphi(n))$ and such that α is continuous or even smooth, on a large subset of N.

No one has yet attempted to give a general bifurcation or catastrophe theory appropriate for functions α which are Lyapunov functions for chain-recurrent sets of vector fields.

EXERCISE 3.1. Let f be a homeomorphism of a compact manifold M and $\Lambda = \Lambda_1 \cup \cdots \cup \Lambda_k$ a disjoint union of closed invariant sets for f. Theorem 2.3 gives necessary and sufficient conditions for the Λ_i to be of the form $K_\alpha^f(\mathbf{M})$ for a filtration $\mathbf{M}$ adapted to f. The conditions are that Λ contain $L(f)$ and that there are no cycles among the Λ_i's. Often Λ will only contain one or the other of $L_+(f)$ or $L_-(f)$. We wish to generalize 2.3 to this case as well; of course, we cannot hope to find a filtration $\mathbf{M}$ with $K^f(\mathbf{M}) = \Lambda$, since $K^f(\mathbf{M})$ contains all of $L(f)$.

We will tackle the case where $L_-(f) \subset \Lambda$; the case where $L_+(f) \subset \Lambda$ follows by replacing f with f^{-1}.

Notice that the proof of Lemma 2.1 readily shows that M is the disjoint union of the $W^u(\Lambda_i)$'s. On the other hand, we cannot say that M is the union of the $W^s(\Lambda_i)$'s.

Define a preorder $>_1$ by $\Lambda_i >_1 \Lambda_j$ if $i \ne j$ and $W^u(\Lambda_i) \cap W^s(\Lambda_j) \ne \emptyset$. Notice that by the first lemma in the proof of 2.3 we have

$$\overline{W^u(\Lambda_i)} \cap W^u(\Lambda_j) = \emptyset \;\Rightarrow\; W^u(\Lambda_i) \cap \Lambda_j \ne \emptyset.$$

Suppose that the preorder $>_1$ has no cycles. We can then extend it to a total order of the Λ_i's; after permuting the indices, one can assume that the total order coincides with that induced by the order of the indices. In this case also, we say that the order of the indices is a filtration ordering.

(1) Show that $\bigcup_{j\leq i} W^u(\Lambda_i)$ is closed.
(2) Let F be a closed set which misses $\bigcup_{k\geq i+1} \Lambda_k$. Show that $\bigcap_{n\geq 0} f^n(F) \subset \bigcup_{j\leq i} W^u(\Lambda_j)$.
(3) Show that there is a closed neighborhood Q of $\bigcup_{j\leq i} W^u(\Lambda_j)$ such that $\bigcap_{n\geq 0} f^n(Q) = \bigcup_{j\leq i} W^u(\Lambda_j)$.
(4) Prove the following theorem:

Theorem 2.3 bis. *Let $\Lambda = \Lambda_1 \cup \cdots \cup \Lambda_k$ be a disjoint union of closed invariant sets for a homeomorphism f. Assume that $L_-(f) \subset \Lambda$; the following two conditions are equivalent:*

(i) *There is a filtration $\mathbf{M}$: $\varnothing = M_0 \subset \cdots \subset M_k \subset M$ adapted to f such that*

$$\bigcap_{n\geq 0} f^n(\mathbf{M}_i) = \bigcup_{j\leq i} W^u(\Lambda_j) \qquad \textit{for all} \quad i = 1, \ldots, k.$$

(ii) *The Λ_i's have no cycles for the preorder $>_1$ and the ordering of indices is a filtration ordering.*

Commentary

Sequences of filtrations are used in [2.6] and [2.2], as well as [1.4] and [1.5], for the topological case; Theorem 3.4 and Proposition 3.5 and 3.6 were formulated in [1.5], Theorem 3.11 in [2.6]. There is a proof of Proposition 3.12 in [2.2] for surfaces and in [2.6] for higher dimensions. Theorem 3.14 does not appear as stated here anywhere in the literature; however, the topological case is treated in [1.4], and in [2.6] one can find the techniques necessary to prove the theorem, but not its statement. The essential ingredient is the smoothing technique in [3.2]. My first contact with these techniques in this setting was [3.1], in which there is a version of Theorem 3.2, stated in terms of flows, which uses Theorem 3.14 and the results of [2.2].

The exercise comes from [2.1].

References

[3.1] Pugh, C. C. and Shub, M., Ω-stability theorem for flows, *Invent. Math.* **11** (1970), 150.

[3.2] Wilson, F. W., Smoothing derivatives of functions and applications, *Trans. Amer. Math. Soc.* **139** (1969), 413.

CHAPTER 4

Hyperbolic Sets

Definition 4.1. Let Λ be an invariant set for a C^r diffeomorphism f of a manifold M. We say that Λ is a *hyperbolic set* for f if there is a continuous splitting of the tangent bundle of M restricted to Λ, TM_Λ, which is Tf invariant:

$$TM_\Lambda = E^s \oplus E^u; \qquad Tf(E^s) = E^s; \qquad Tf(E^u) = E^u;$$

and for which there are constants c and λ, $c > 0$ and $0 < \lambda < 1$, such that

$$\|Tf^n|_{E^s}\| < c\lambda^n, \qquad n \geq 0,$$

$$\|Tf^{-n}|_{E^u}\| < c\lambda^n, \qquad n \geq 0.$$

REMARK. This condition is independent of the metric on M, since if $\|\ \|_1$ and $\|\ \|_2$ are two equivalent norms of TM_x, there are strictly positive constants c_1' and c_2' such that

$$c_1'\|\ \|_2 \leq \|\ \|_1 \leq c_2'\|\ \|_2.$$

If M is compact we can make a similar estimate globally on TM:

$$\exists c_1 > 0, \quad c_2 > 0 \qquad \text{such that} \quad c_1\|\ \|_2 \leq \|\ \|_1 \leq c_2\|\ \|_2.$$

Now the independence of the notion of hyperbolicity from the choice of metric follows since

$$\|Tf^n(v)\|_1 \leq c_1\lambda^n\|v\|_1, \qquad v \in E^s, \qquad n \geq 0,$$

implies

$$c_1\|Tf^n(v)\|_2 \leq c_2 c\lambda^n\|v\|_2, \qquad v \in E^s, \qquad n \geq 0,$$

so

$$\|Tf^n(v)\|_2 \leq (c_2/c_1)c\lambda^n\|v\|_2, \qquad v \in E^s, \qquad n \geq 0,$$

and we have, finally,

$$\|Tf^n|_{E^s}\|_2 \leq (c \cdot c_2/c_1)\lambda^n.$$

Proposition 4.2. *Suppose that $\Lambda \subset M$ is a hyperbolic set for a C^r diffeomorphism f of M. Then there is a C^∞ metric on M and a constant σ, $0 < \sigma < 1$ such that:*

$$\|Tf|_{E^s}\| < \sigma \quad \text{and} \quad \|(Tf^{-1})|_{E^u}\| < \sigma.$$

PROOF. Take an arbitrary metric $v \mapsto |v|$, $v \in TM$, on M. We know that we can find constants $0 < c$, $0 < \lambda < 1$ such that

$$|Tf^k|_{E^s}| < c\lambda^k; \qquad |Tf^{-k}|_{E^u}| < c\lambda^k, \qquad k \geq 0.$$

Choose an n so large that $(c\lambda^n)$ is much less than 1 and define a new metric on E^s by

$$\|v\|^2 = \sum_{j=0}^{n-1} |(Tf)^j(v)|^2, \qquad v \in E^s.$$

We construct an analogous metric on E^u using Tf^{-1} instead of Tf and put the orthogonal sum metric on TM_Λ; then we extend this any way we choose to all of TM. This is not yet a C^∞ metric because the function $\| \ \|^2: TM \to \mathbb{R}$ is not necessarily C^∞ since E^u and E^s need only be continuous subbundles. Nevertheless, $\| \ \|$ is a norm on each fiber of TM and $\| \ \|^2: TM \to \mathbb{R}$ is continuous.

For all v in E^s, we have

$$\|Tf(v)\|^2 = \sum_{j=0}^{n-1} |Tf^j(Tf(v))|^2 = \sum_{j=0}^{n-1} |Tf^{j+1}(v)|^2$$

$$= \sum_{j=1}^{n} |Tf^j(v)|^2 = \sum_{j=0}^{n-1} |Tf^j(v)|^2 + |Tf^n(v)|^2 - |v|^2.$$

However, since $|Tf^n(v)|$ is less than or equal to $c\lambda^n|v|$, we have

$$\|Tf(v)\|^2 \leq \|v\|^2 - |v|^2(1 - (c\lambda^n)^2).$$

We can assume that c is greater than or equal to 1 (otherwise there would have been nothing to prove). We then have

$$\|v\|^2 = \sum_{j=0}^{n-1} |Tf^j(v)|^2 \leq |v|^2 + c^2\lambda^2|v|^2 + \cdots + c^2\lambda^{2(n-1)}|v|^2$$

and thus

$$\|v\|^2 \leq c^2 n|v|^2 \quad \text{so} \quad (\|v\|^2/c^2 n) \leq |v|^2.$$

From this we get the inequalities

$$\|Tf(v)\|^2 \leq \|v\|^2 - (\|v\|^2/c^2 n)(1 - (c\lambda^n)^2)$$

and thus

$$\|Tf(v)\|^2 \leq (1 - (1 - (c\lambda^n)^2)/c^2 n)\|v\|^2.$$

Therefore, setting $\sigma = \sqrt{1 - (1 - (c\lambda^n)^2)/c^2 n}$, we have

$$\|Tf(v)\| \leq \sigma, \quad \text{with} \quad 0 < \sigma < 1.$$

The same proof applies to $Tf^{-1}|_{E^u}$. To conclude we need only approximate by a C^∞ metric on TM, so closely that $Tf|_{E^s}$ and $Tf|_{E^u}$ remain close in norm. □

Example 4.3 (A Hyperbolic Fixed Point). Suppose that Λ reduces to a point p; we have $f(p) = p$ and $T_p f$ is a linear isomorphism of $T_p M$ onto itself. There will be a hyperbolic splitting of $T_p M$ in this case if and only if all the eigenvalues of $T_p f$ have modulus different from 1.

PROOF. Exercise; let E^s be the generalized eigenspace corresponding to the eigenvalues of modulus less than 1 and E^u the generalized eigenspace corresponding to those of modulus greater than 1. □

When M is of dimension 2, we have the three families of pictures shown in Figure 4.1.

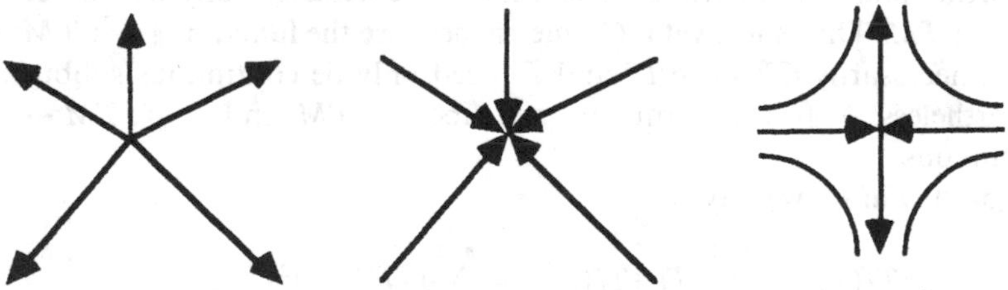

Figure 4.1. In (a) we have a *source*; both eigenvalues are outside the unit circle and $E^u = \mathbb{R}^2$, $E^s = \{0\}$. In (b) we have a *sink*; both eigenvalues are inside the unit circle and $E^u = \{0\}$, $E^s = \mathbb{R}^2$. In (c) we have a *saddle*; one eigenvalue inside and one outside the unit circle. Here E^s is the x axis and E^u the y axis. Note that every orbit not contained in E^u is asymptotic to E^s so that these orbits are truly "hyperbolic".

Example 4.4 (A Hyperbolic Periodic Orbit). Suppose that Λ is a periodic orbit; $\Lambda = \{p_1, \ldots, p_k\}$; $f(p_i) = p_{i+1}$ (mod k). Then we have:

$$TM_\Lambda = (T_{p_1}M \cup \cdots \cup T_{p_k}M).$$

A hyperbolic splitting of TM_Λ is thus a Tf invariant splitting of the tangent of M over points in the orbit Λ:

$$T_{p_i}M = E^s_{p_i} \oplus E^u_{p_i}$$

$$Tf(E^s_{p_i}) = E^s_{p_{i+1}} \quad (\text{mod } k); \qquad Tf(E^u_{p_i}) = E^u_{p_{i+1}} \quad (\text{mod } k).$$

REMARKS. (1) Tf^k is an automorphism of each $T_{p_i}M$.

(2) There is a hyperbolic splitting of TM_Λ if and only if each p_i is a hyperbolic fixed point of f^k.

Definition 4.5. We say that a C^1 diffeomorphism f of M is *Anosov* if M is a hyperbolic set for f.

Example 4.6. Let A be the integral matrix $\begin{bmatrix} 2 & 1 \\ 1 & 1 \end{bmatrix}$. The determinant of A is 1, which implies that A is an automorphism of the lattice $\mathbb{Z}^2 \subset \mathbb{R}^2$. The matrix A thus acts on the quotient torus $T^2 = \mathbb{R}^2/\mathbb{Z}^2$ so the following diagram commutes:

$$\begin{array}{ccc} \mathbb{R} & \xrightarrow{A} & \mathbb{R}^2 \\ \downarrow & & \downarrow \\ T^2 = \mathbb{R}^2/\mathbb{Z}^2 & \xrightarrow{\bar{A}} & \mathbb{R}^2/\mathbb{Z}^2 = T^2. \end{array}$$

Moreover, since the eigenvalues of A are $(3 \pm \sqrt{5})/2$, T^2 is hyperbolic for f.

More generally, a matrix A in $SL_n(\mathbb{Z})$ gives rise to an automorphism of the torus $T^n = \mathbb{R}^n/\mathbb{Z}^n$. T^n is a hyperbolic invariant set for $\bar{A}$ if and only if A has no eigenvalues of modulus 1. In this case $\bar{A}$ is an Anosov diffeomorphism of T^n.

Example 4.7 (Smale's Horseshoe). In order to construct a diffeomorphism of the sphere S^2 (or any other surface) having a nontrival infinite nonwandering set, we first construct a mapping of the square $I \times I$ into the plane and determine its nonwandering set.

We begin by stretching the square R into a bar (Figure 4.2) by means of a linear map, then bending this pliant bar into a horseshoe and placing it back on R. We assume that the restrictions of f to B and D are affine maps which stretch vertically and contract horizontally. We now extend this to a mapping of the disc D^2 into itself by adding two half-discs to R above and below, as shown in Figure 4.3. The image under f of the lower half-disc Δ is contained

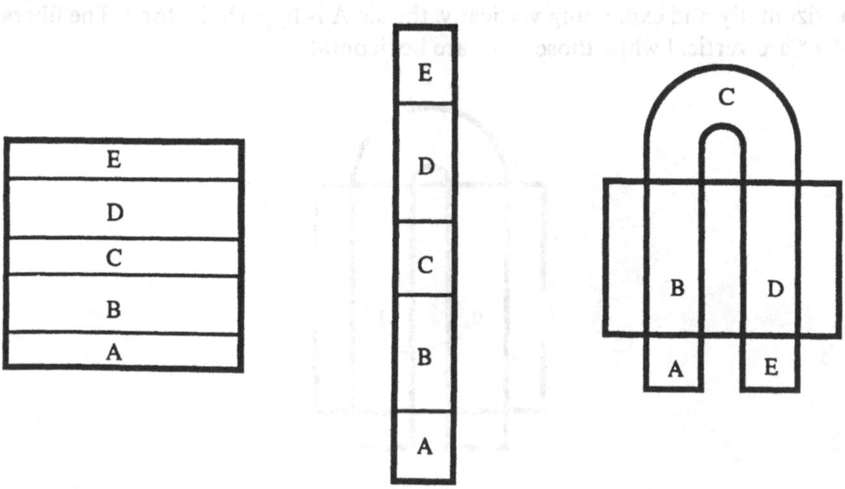

Figure 4.2.

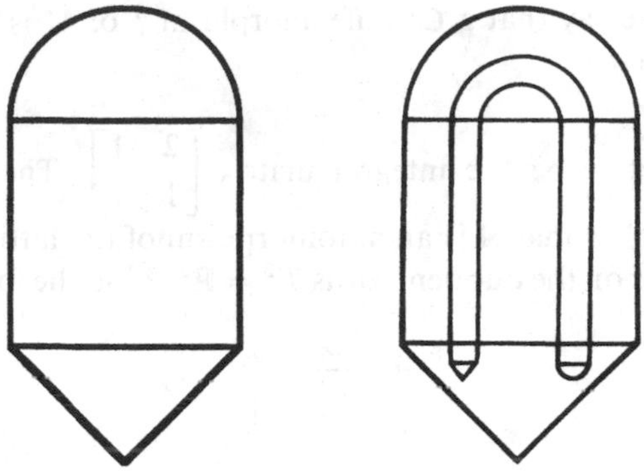

Figure 4.3.

in Δ and the extension of f to Δ may be chosen to be contractive. Thus f restricted to Δ has a unique fixed point p_1 which is all of the nonwandering set that lies in Δ and it is a sink. By adding a source at the antipodal point of S^2, p_2, we can extend f to a diffeomorphism of S^2 which we persist in calling f. Of course, we are ignoring the trifling fact that Δ has corners. We have a filtration $\mathbf{M}$ adapted to f, $\emptyset = M_0 \subset \Delta = M_1 \subset D^2 = M_2 \subset M^3 = S^2$. Since the restriction of f to Δ is a contraction, we have $K_1(\mathbf{M}) = p_1$. Similarly, since the restriction of f to $S^2 - D^2$ is a dilation, we have $K_3(\mathbf{M}) = p_2$. By definition, $K_2(\mathbf{M}) = \bigcap_{n \in \mathbb{Z}} f^n(F)$. Setting $\Lambda = K_2(M)$, we have

$$\Omega(f) = p_1 \cup p_2 \cup (\Omega(f) \cap R) \quad \text{and} \quad \Omega(f) \cap R \subset \Lambda.$$

We will prove $\Omega(f) \cap R = \Lambda$, which shows that $\mathbf{M}$ is a fine filtration for f and we will analyze Λ.

The set $R \cap f(R)$ (Figure 4.4) has two connected components which we will label "0" and "1". Since we have assumed f to be affine on B and D, contracting horizontally and expanding vertically, the set Λ is hyperbolic for f. The fibers of E^u are vertical while those of E^s are horizontal.

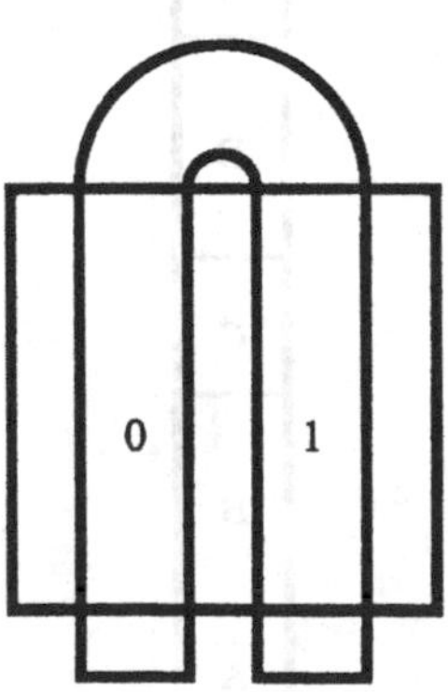

Figure 4.4.

To each point x of Λ we associate an infinite sequence of 0's and 1's, $\{a_n\}$ as follows:

$$a_n(x) = 0 \quad \text{if} \quad f^n(x) \in 0,$$
$$a_n(x) = 1 \quad \text{if} \quad f^n(x) \in 1.$$

This defines a mapping

$$\Phi\colon \Lambda \to \prod_{-\infty}^{+\infty} \{0, 1\},$$
$$x \mapsto \{a_n(x)\}.$$

We denote by $\Sigma(2)$ the space $\prod_{-\infty}^{+\infty} \{0, 1\}$ with the product topology. The space $\Sigma(2)$ is compact, perfect, and totally disconnected, that is to say, a Cantor set. An open basis for the topology of $\Sigma(2)$ is given by the open sets

$$C(\varepsilon_1, \ldots, \varepsilon_k; i_1, \ldots, i_k); \qquad \varepsilon_j \in \{0, 1\},$$

which consists of those sequences $\{a_n\}$ with $a_{i_1} = \varepsilon_1, \ldots, a_{i_k} = \varepsilon_k$. There is a natural automorphism of $\Sigma(2)$, the shift σ given by $(\sigma\{a_n\})_k = a_{k+1}$. Thus the mapping Φ above makes the following diagram commute:

$$\begin{array}{ccc} \Lambda & \xrightarrow{f|_\Lambda} & \Lambda \\ \Phi \downarrow & & \downarrow \Phi \\ \Sigma(2) & \xrightarrow{\sigma} & \Sigma(2). \end{array}$$

Theorem 4.8. *The mapping $f|_{\Omega(f)} \cap R$ is topologically conjugate to the shift σ; more precisely, Φ is a homeomorphism and $\Lambda = \Omega(f) \cap R$.*

PROOF. We already know that $\Omega(f) \cap R$ is contained in Λ. The reverse inclusion is a consequence of the rest of the theorem since every point of $\Sigma(2)$ is nonwandering for the shift σ (cf. Exercise 4.11).

(1) *The mapping Φ is continuous.* Let x be a point of Λ and U a neighborhood of $\Phi(x)$. Choose an open cylinder: $\mathrm{Cyl}(\Phi(x), N) = \{\{y_n\} | y_i = (\Phi(x))_i \text{ for } |i| \leq N\}$ contained in U. For every i with absolute value less than or equal to N, choose a ball of positive radius δ_i: $B_{\delta_i}(f^i(x))$ centered at $f^i(x)$ which misses the components of $f(R) \cap R$ which do not contain $f^i(x)$. For all i between $-N$ and N, $(\Lambda \cap f^{-i}(B_{\delta_i}(f^i(x))))$ is a neighborhood of x in Λ. The same is true for finite intersections of such neighborhoods, that is,

$$B_N(x) = \bigcap_{|i| \leq N} (\Lambda \cap f^{-i}(B_{\delta_i}(f^i(x))))$$

is also a neighborhood of x in Λ. By construction, $\Phi(B_N(x))$ is contained in the cylinder $\mathrm{Cyl}(\Phi(x), N)$, so Φ is continuous.

(2) *The mapping Φ is injective.* Recall the construction of f, illustrated in Figure 4.5. The set $f^{-1}(R) \cap R$ is the union of two horizontal bands B and D.

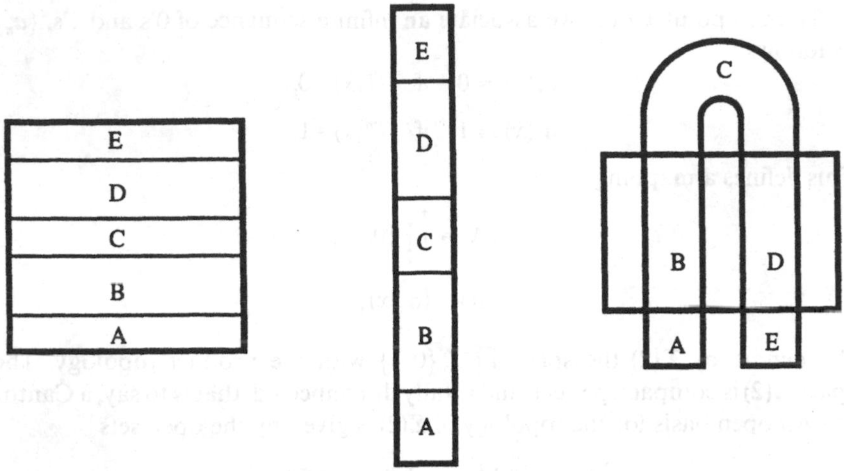

Figure 4.5.

Let x and y be two points in $\Lambda = (\bigcap_{k \in Z} f^k R)$, such that $\Phi(x) = \Phi(y)$. We will first show that if the positive parts of the sequences coincide, $(\Phi(x))_i = (\Phi(y))_i$; $\forall i \geq 0$, then the ordinates of x and y are equal. In fact, $f^i(x)$ and $f^i(y)$ will belong to the same vertical band of $f(R) \cap R$, and also to the same horizontal band of $f^{-1}(R) \cap R$, since $f^{i+1}(x)$ and $f^{i+1}(y)$ belong to the same vertical band. But because $f^i(x)$ and $f^i(y)$ belong to the same horizontal band of $f^{-1}(R) \cap R$, we will have

$$|\text{ordinate}(f^{i+1}(y)) - \text{ordinate}(f^{i+1}(x))| = \lambda |\text{ord}(f^i(x)) - \text{ord}(f^i(y))|,$$

where λ is the vertical stretch of f on $B \cup D$ and is strictly greater than 1. We thus have, for all n,

$$|\text{ord}(f^n(x)) - \text{ord}(f^n(y))| = \lambda^n |\text{ord}(x) - \text{ord}(y)|$$

which, if $f^n(x)$ and $f^n(y)$ are to remain in R, is impossible unless x and y have the same ordinate.

By analyzing the past of the points x and y, one can also show that if the negative parts of the sequences $(\Phi(x))_i$ and $(\Phi(y))_i$ coincide, then the two points have the same abscissa.

(3) *The mapping Φ is surjective.* We denote by I_0 and I_1 the two components, "0" and "1", of $f(R) \cap R$. The image of each component of $f(R) \cap R$ is a narrower horseshoe which traverses both components of $f(R) \cap R$ (Figure 4.6).

We thus obtain four vertical bands:

$$(0, 0) = I_0 \cap f(I_0); \qquad (0, 1) = I_0 \cap f(I_1);$$

$$(1, 0) = I_1 \cap f(I_0); \qquad (1, 1) = I_1 \cap f(I_1);$$

Let $\alpha = (a_0, a_1, \ldots, a_N)$ be a finite sequence of 0's and 1's and assume that

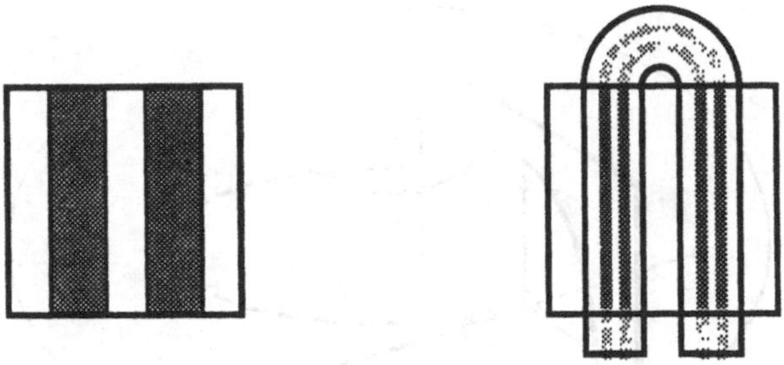

Figure 4.6.

$I_\alpha = I_{a_0} \cap f(I_{a_1}) \cap \cdots \cap f^N(I_{a_N})$ is a nonempty vertical band contained either in I_0 or I_1. Thus we have shown that

$$I_0 \cap f(I_\alpha) = I_0 \cap f(I_{a_1}) \cap \cdots \cap f^{N+1}(I_{a_N})$$

and

$$I_1 \cap f(I_\alpha) = I_1 \cap f(I_{a_1}) \cap \cdots \cap f^{N+1}(I_{a_N})$$

are two nonempty vertical bands which traverse R.

By induction, to every finite sequence $a_0, \ldots, a_n$ there corresponds such a nonempty vertical band:

$$I_{a_0} \cap f(I_{a_1}) \cap \cdots \cap f^n(I_{a_N}) \neq \emptyset.$$

Now let $\alpha = \{a_i\}_{i \in \mathbb{Z}}$. We wish to show that the intersection $I_\alpha = \bigcap_{i=-\infty}^{+\infty} f^i(I_{a_i})$ is nonempty. In fact, if x belongs to I_α, $f^i(x)$ belongs to I_{a_i}, for any i and $\Phi(x) = \{a_i\}_{i \in \mathbb{Z}}$. For this, it suffices to show that all intersections corresponding to finite sequences are nonempty. That is, if all the sets $I_{\alpha,N} = \bigcap_{-N}^{+N} f^i(I_{a_i})$ are nonempty, they form a nested sequence of nonempty closed subsets of the compact set Λ; hence the intersection $I_\alpha = \bigcap_N I_{\alpha,N}$ is also nonempty.

We have seen that for any finite sequence $b_0, \ldots, b_N$ of 0's and 1's, $I_{b_0} \cap f(I_{b_1}) \cap \cdots \cap f^n(I_{b_N})$ is a nonempty vertical band. Thus, given a finite sequence $a_{-N}, \ldots, a_0, \ldots, a_N$ the intersection $I_{a_{-N}} \cap f(I_{a_{1-N}}) \cap \cdots \cap f^N(I_{a_0}) \cap \cdots \cap f^{2N}(I_{a_N})$ is nonempty. Therefore, the image under f^{-N} of this set,

$$f^{-N}(I_{a_{-N}}) \cap \cdots \cap I_{a_0} \cap \cdots \cap f^N(I_{a_N})$$

is also nonempty, and the theorem is proven. □

Example 4.9 (The Solenoid). We begin with a solid torus P (Figure 4.7) embedded in three-dimensional Euclidean space, $\mathbb{R}^3$. We represent points of P by means of coordinates $(\theta, \bar{r}, s)$; θ is an angle, $\theta \in S^1$; $\bar{r}$ and s are real numbers between -1 and 1 such that $\bar{r}^2 + s^2 \leq 1$. The point x with coordinates $(\theta, \bar{r}, s)$ belongs to the plane orthogonal to the core of the torus (which is a horizontal circle with radius 2 thought of as S^1) through the point θ in S^1 having position

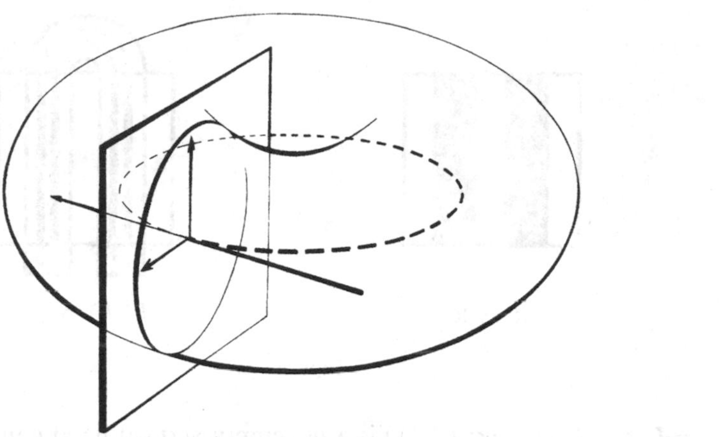

Figure 4.7. Here T is tangent to the core of the torus, e_1 lies in the equatorial plane of the torus and (T, e_1, e_2) form a positively oriented orthogonal frame.

$(\bar{r}, s)$ relative to the frame (e_1, e_2) in this plane (cf. Figure 4.7). Notice that the points $(\theta, \bar{r}, 0)$ comprise an annulus of radii 1 and 3 containing the core of P.

We define a mapping $f: P \to P$ by

$$f(\theta, \bar{r}, s) = (2\theta, \varepsilon_1 \cos\theta + \varepsilon_2 \bar{r}, \varepsilon_1 \sin\theta + \varepsilon_2 s),$$

where ε_1 and ε_2 are two small positive constants, $0 < \varepsilon_2 < \varepsilon_1 < \frac{1}{2}$.

The image $f(P)$ is contained in P. A disc perpendicular to the core at θ is mapped into a disc perpendicular to the core at 2θ (Figure 4.8). It is easy to see that the conditions $\overline{r^2} + s^2 \leq 1$, $\varepsilon_2 < \varepsilon_1 < \frac{1}{2}$ guarantee that $f: P \to P$ is an embedding. In fact, let D_θ be the section of P by a plane perpendicular to the core at θ. The images $f(D(\theta/2))$ and $f(D(\theta/2 + \pi))$ are both contained in

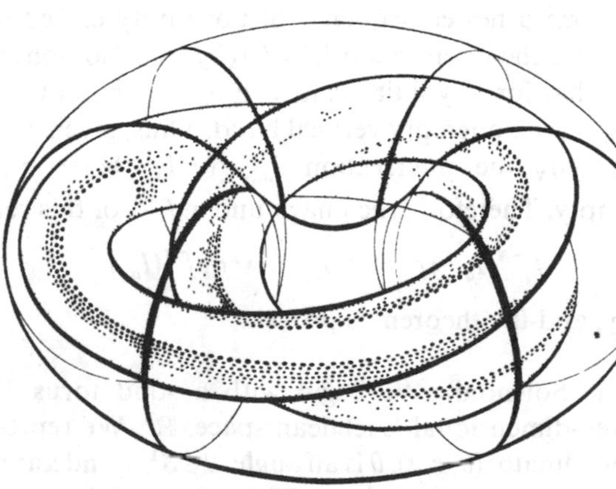

Figure 4.8.

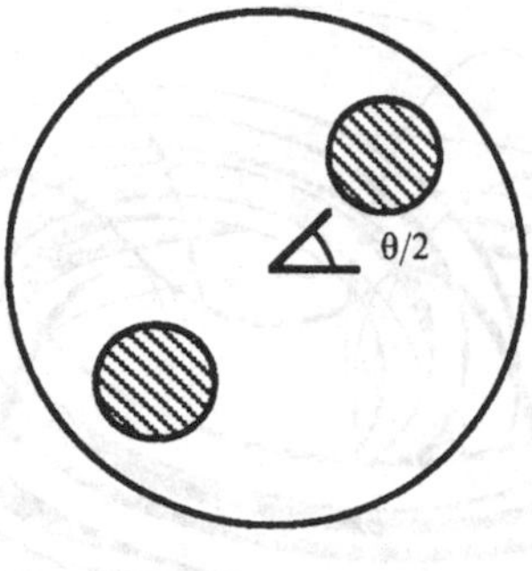

Figure 4.9.

$D(\theta)$ (Figure 4.9) and are similar to $D(\theta)$ with a factor of similarity of ε_2. The coordinates of the centers of these discs in the plane of $D(\theta)$ are

$$\left(\varepsilon_1 \cos\frac{\theta}{2}, \varepsilon_1 \sin\frac{\theta}{2}\right)$$

and

$$\left(\varepsilon_1 \cos\left(\frac{\theta}{2}+\pi\right), \varepsilon_1 \sin\left(\frac{\theta}{2}+\pi\right)\right) = -\left(\varepsilon_1 \cos\frac{\theta}{2}, \varepsilon_1 \sin\frac{\theta}{2}\right).$$

These two centers are separated by a distance of $2\varepsilon_1$ and the image discs have radii ε_2, so f is indeed an embedding.

A diagram of the intersection of $f^2(P)$ with $D(\theta)$ is shown in Figure 4.10. A curve $f(\psi, \overline{r_0}, s_0)$, where ψ is a variable angle, which is a fourfold cover of the core, passes through each point of $f^2(P) \cap D(\theta)$ with coordinates $(\overline{r_0}, s_0)$. The curves are transverse to each disk $D(\psi)$.

Altogether, $f^2(P)$ looks approximately as shown in Figure 4.11. Consider the discs $D(\theta) = \{ (\theta, \overline{r}, s) | \overline{r^2} + s^2 \leq 1\}$; we can introduce polar coordinates (φ, ρ) in them. The intersection $f(P) \cap D(\theta)$ is the union of two discs with centers $(\varphi = \theta/2, \varepsilon_1)$ and $(\varphi = \theta/2 + \pi, \varepsilon_1)$.

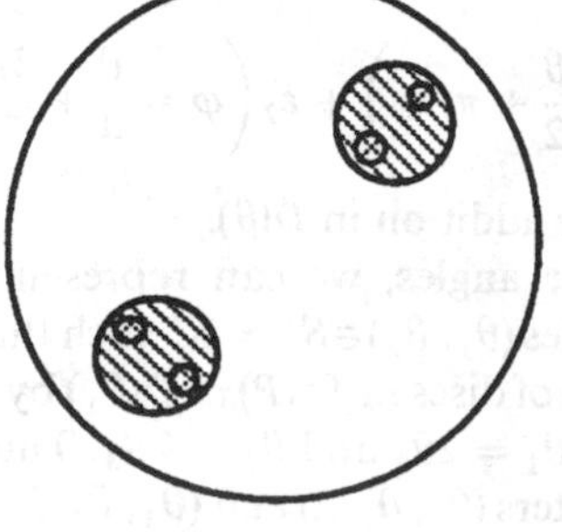

Figure 4.10.

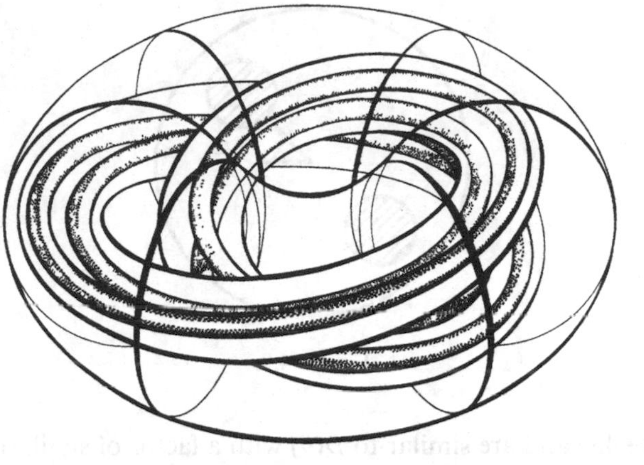

Figure 4.11.

We now describe $f^2(P) \cap D(\theta)$. We begin by determining the intersection $f(P) \cap D(\theta/2)$ (resp. $f(P) \cap D(\theta/2 + \pi)$). In each of these discs the image of $f(P)$ is the union of two discs centered, respectively, at

$$\left(\varphi = \frac{\theta}{4}, \varepsilon_1\right); \quad \left(\varphi = \frac{\theta}{4} + \pi, \varepsilon_1\right)$$

$$\left(\text{resp.} \left(\varphi = \frac{\theta}{4} + \frac{\pi}{2}, \varepsilon_1\right), \left(\varphi = \frac{\theta}{4} + \frac{3\pi}{2}, \varepsilon_1\right)\right).$$

Under f these discs are sent to four discs in $D(\theta)$ of radius ε_2^2 centered at

$$\left(\varphi = \frac{\theta}{2}, \varepsilon_1\right) + \varepsilon_2\left(\varphi = \frac{\theta}{4}, \varepsilon_1\right);$$

$$\left(\varphi = \frac{\theta}{2}, \varepsilon_1\right) + \varepsilon_2\left(\varphi = \frac{\theta}{4} + \pi, \varepsilon_1\right);$$

$$\left(\varphi = \frac{\theta}{2} + \pi, \varepsilon_1\right) + \varepsilon_2\left(\varphi = \frac{\theta}{4} + \frac{\pi}{2}, \varepsilon_1\right);$$

$$\left(\varphi = \frac{\theta}{2} + \pi, \varepsilon_1\right) + \varepsilon_2\left(\varphi = \frac{\theta}{4} + \frac{3\pi}{2}, \varepsilon_1\right);$$

where $+$ refers to vector addition in $D(\theta)$.

By only retaining the angles, we can represent the center of a disc in $f(P) \cap D(\theta_1)$ by two angles $(\theta_1, \theta_2) \in S^1 \times S^1$ such that $2\theta_2 = \theta_1$. Similarly, we can represent the centers of discs in $f^2(P) \cap D(\theta_1)$ by three angles $(\theta_1, \theta_2, \theta_3) \in S^1 \times S^1 \times S^1$ such that $\theta_1 = 2\theta_2$ and $\theta_2 = 2\theta_3$. Thus, if we fix θ_1 we have in $D(\theta_1)$ two discs with centers $(\theta_1, \theta_1/2)$ and $(\theta_1, \theta_1/2 + \pi)$ and radii ε_2 and four discs with centers $(\theta_1, \theta_1/2, \theta_1/4)$, $(\theta_1, \theta_1/2, \theta_1/4 + \pi)$, $(\theta_1, \theta_1/2 + \pi, \theta_1/4 + \pi/2)$, $(\theta_1, \theta_1/2 + \pi, \theta_1/4 + 3\pi/2)$ and radii ε_2^2 (cf. Figure 4.10). Each disc in

$f^n(P) \cap D(\theta)$ thus contains two discs in $f^{n+1}(P) \cap D(\theta)$ with radii ε_2 times smaller.

Continuing indefinitely we obtain an infinite sequence

$$(\theta_1, \theta_2, \ldots, \theta_k, \theta_{k+1}, \ldots) \in S^1 \times S^1 \times \cdots \times S^1 \times S^1 \times \cdots$$

satisfying $2\theta_{k+1} = \theta_k$ for all k strictly greater than 1. This sequence determines an infinite sequence of nested discs whose radii tend to 0, and thus a single point of $\bigcap_{n \in \mathbb{N}} f^n(P)$.

We can now identify points of $\bigcap_{n \in \mathbb{N}} f^n(P)$ with sequences as above; such sets are called solenoids. Each disc $D(\theta)$ cuts the solenoid in a Cantor set and the solenoid is locally the product of a Cantor set and an interval as Figure 4.10 suggests.

The set of sequences $(\theta_1, \theta_2, \ldots) \in S^1 \times S_1 \times \cdots$ such that $\theta_k = 2\theta_{k+1}$ is also known as the projective limit of the sequence $S^1 \overset{\times 2}{\leftarrow} S^1 \overset{\times 2}{\leftarrow} \ldots, 2\theta \leftarrow \theta \ldots$.

The restriction of f to the solenoid, which is invariant under f, can be described in these terms as follows:

$$f\colon (\theta_1, \theta_2, \ldots, \theta_k, \theta_{k+1}, \ldots) \mapsto (2\theta_1, \theta_1, \ldots, \theta_{k-1}, \theta_k, \ldots).$$

Thus f induces a homeomprohism of the solenoid, which is compact in the projective limit of topology.

The solenoid is a hyperbolic invariant set. In fact, the tangent bundle TP splits as $TP = E^s \oplus F^u$, where E^s is tangent to the meridional discs and F^u is tangent to the parallels $S^1 \times \text{pt.} \subset S^1 \times D^2$). The image under f of a meridional disc being a meridional disc, we see that E^s is Tf invariant and, further,

$$\| Tf|_{E^s} \| < \varepsilon_2.$$

On the other hand, the f image of a parallel cuts the meridional discs at an angle α close to $\pi/2$.

Let p be the projection of TP onto F^u. One can check that, for all vectors v in F^u, we have

$$|p \circ Tf(v)| = 2|v|.$$

Since F^u is almost invariant, we will see, in the exercises at the end of Chapter 7, that there is a hyperbolic splitting of the restriction of TP to the solenoid.

EXERCISE 4.1. In Definition 4.1 we have implicitly assumed that E^s and E^u are subbundles (locally trivial) of TM_Λ. Show that if we only assume that at each point x of Λ there is a splitting of $T_xM = E^s_x \oplus E^u_x$ satisfying the properties of Definition 4.1, then the subspaces E^s_x (resp. E^u_x) form a subbundle of TM_Λ. *Hint*: Show that E^s_x depends continuously on x.

EXERCISE 4.2. Let σ be the shift $\sigma\colon \Sigma(2) \to \Sigma(2)$ as before. Show that:

(1) $\overline{\mathrm{Per}(\sigma)} = \Sigma(2)$;
(2) $\Omega(\sigma) = \Sigma(2)$;
(3) the number of periodic points of period n of σ, $N_n(\sigma) = 2^n$.

Commentary

My examples and definitions come from [1.16], there the reader will find the basic facts and motivations which will help in reading this chapter. Proposition 4.2 is from [4.2] and was first stated in [4.4].

It seems to me, that at one time, many people wondered if there were a stable diffeomorphism with infinitely many periodic orbits. Levinson [4.3], however, gave an example of this. Smale [4.6] constructed the horseshoe while considering Levinson's example and linked it to Poincaré's homoclinic points. Thom first noticed Example 4.6, which is sometimes called the Thom diffeomorphism. Anosov [4.1] showed that what are now called Anosov diffeomorphisms are structurally stable and formulated the global condition of hyperbolicity. Smale [1.16] defined hyperbolic sets, then Ω-stability and generalized Anosov's theorem to the case of Ω-stability.

I particularly like the example of the solenoid. Smale found it after I studied, in my thesis [4.5], the mapping from S^1 to S^1 given by $z \to z^2$, which I showed to be structurally stable, as are all expanding maps.

Williams [4.7] generalized these results to expanding attractors. I have taken, in this chapter, the point of view of Williams.

References

[4.1] Anosov, D. V., Geodesic flows on compact manifolds of negative curvature, *Trudy Mat. Inst. Steklov* **90** (1967); *Proc. Steklov Inst. Math.* (transl.) (1969).

[4.2] Hirsch, M. W. and Pugh, C. C., Stable manifolds and hyperbolic sets, in *Global Analysis*, Vol. XIV (Proceedings of Symposia in Pure Mathematics), American Mathematical Society, Providence, R. I., 1970, p.133.

[4.3] Levinson, N., A second-order differential equation with singular solutions, *Ann. of Math.* **50** (1949), 126.

[4.4] Mather, J., Characterization of Anosov diffeomorphisms, *Nederl. Akad. Wetensch. Indag. Math.* **30** (1968), 479.

[4.5] Shub, M., Endomorphisms of compact differentiable manifolds, *Amer. J. Math.* **91** (1969), 175.

[4.6] Smale, S., Diffeomorphisms with many periodic points, in *Differential and Combinatorial Topology, a Symposium in Honor of M. Morse*, S. S. Cairns (Ed.), Princeton University Press, Princeton, N. J., 1965, p. 63.

[4.7] Williams, R. F., Expanding attractors, *Institut Hautes Études Sci. Publ. Math.* **43** (1974), 169.

CHAPTER 5

Stable Manifolds

We have begun to study hyperbolic invariant sets. Before we continue we must generalize our definitions in the case of a fixed point, to allow us to work in an arbitrary Banach space.

Definition 5.1. Consider an endomorphism, $T\colon E \to E$ that is a continuous linear map, T of a Banach space E to itself. We say that T is *hyperbolic* if and only if E has a T invariant direct sum decomposition $E = E_1 \oplus E_2$, for which there exist constants $c > 0$ and $\lambda < 1$ such that:

(1) The restriction T_1 of T to E_1 is an expansion:

$$\forall n \leq 0, \qquad \|T_1^n\| \leq c\lambda^{-n}.$$

(2) The restriction T_2 of T to E_2 is a contraction:

$$\forall n \geq 0, \qquad \|T_2^n\| \leq c\lambda^n.$$

Notice that Proposition 4.2 allows us to replace the given norm on E by an adapted norm for which we can take $c = 1$ above.

Recall that a map g between two metric spaces is called Lipschitz if there is a constant k such that

$$\forall x, y \qquad d(g(x), g(y)) \leq kd(x, y).$$

The least such k is called the Lipschitz constant of g, $\mathrm{Lip}(g)$. We denote by $E_i(r)$ the closed ball of radius r about this origin in E_i.

Theorem 5.2 (The Local Unstable Manifold Theorem for a Point). *Let $T\colon E \to E$ be a hyperbolic automorphism of the Banach space E with splitting $E = E_1 \oplus E_2 = E_1 \times E_2$ and suppose that the norm $\|\ \|$ is adapted so we can* *find a λ between 0 and 1 such that*

$$\|T|_{E_2}\| < \lambda \quad \text{and} \quad \|T^{-1}|_{E_1}\| < \lambda.$$

Then there is an $\varepsilon > 0$, which depends only on λ, and constant $\delta = \delta(\lambda, \varepsilon, r)$ such that for all Lipschitz maps $f\colon E_1(r) \times E_2(r) \to E$, with $\|f(0)\| < \delta$ and $\mathrm{Lip}(f - T) < \varepsilon$, there is a map $g\colon E_1(r) \to E_2(r)$ whose graph gives a local unstable manifold for f. More precisely, the following six conditions hold:

(1) *g is Lipschitz with $\mathrm{Lip}(g) \leq 1$. Moreover, the restriction of f^{-1} to the graph of g is contracting and thus has a fixed point p on the graph of g.*
(2) *The graph of g is equal to $\bigcap_{n=0}^{\infty} f^n(E_1(r) \times E_2(r))$. (This intersection is the local unstable set of p, $W^{\mathrm{u}}_{\mathrm{loc}}(p)$; since it is in the graph of g it is a manifold as well, hence the local unstable manifold.)*
(3) *g is C^k if f is.*
(4) *If f is C^1 with $f(0) = 0$, $Df(0) = T$, then the graph of g is tangent to E_1 at 0.*
(5) *If $f(0) = 0$ and f is invertible, the graph g consists of those points in $E_1(r) \times E_2(r)$ whose backwards iterates tend to 0; the graph of g is $W^{\mathrm{u}}_r(0)$.*
(6) *If $f(0) = 0$, a point x belongs to the graph of g if and only if there is a sequence x_n, $n \geq 0$, in $E_1(r) \times E_2(r)$, tending to 0 such that $f^n(x_n) = x$.*

We obtain the local stable manifold by exchanging T and T^{-1}, E_1 and E_2.

The proof is so long that it occupies the rest of the chapter. We begin by fixing some notations:

$$T_i = T|_{E_i}, \qquad p_i = \text{projection of } E \text{ onto } E_i, \qquad f_i = p_i \circ f, \qquad i = 1, 2.$$

We will use the more convenient, but equivalent, box norm $\|\ \|_{\mathrm{box}} = \sup(\|\ \|_{E_1}, \|\ \|_{E_2})$; i.e., $\|x\| = \sup(\|p_1(x)\|, \|p_2(x)\|)$.

I. The Case of f Merely Lipschitz: The Graph Transform

Definition 5.3. Suppose that we have a $\sigma\colon E_1(r) \to E_2(r)$ for which $f_1 \circ (\mathrm{id}, \sigma)$ is injective and overflows $E_1(r)$, that is $E_1(r) \subset f_1 \circ (\mathrm{id}, \sigma)(E_1(r))$. We define the function $\Gamma_f(\sigma)\colon E_1(r) \to E_2(r)$ as follows:

$$\Gamma_f(\sigma) = f_2 \circ (\mathrm{id}, \sigma) \circ [f_1 \circ (\mathrm{id}, \sigma)]^{-1}|_{E_1(r)}.$$

This is illustrated in Figure 5.1.

This may sound bizarre, but notice that the graph of $\Gamma_f(\sigma)$ is the intersection of f(graph of σ) with $E_1(r) \times E_2(r)$, so Γ_f deserves to be called the (local, nonlinear) graph transform. Notice that the unstable manifold of T is E_1 which is the only graph invariant under Γ_T, so there is hope of finding the unstable manifold of f as a fixed point of Γ_f.

Let $\mathrm{Lip}_1(E_1(r), E_2(r))$ be the set of Lipschitz functions with constant less

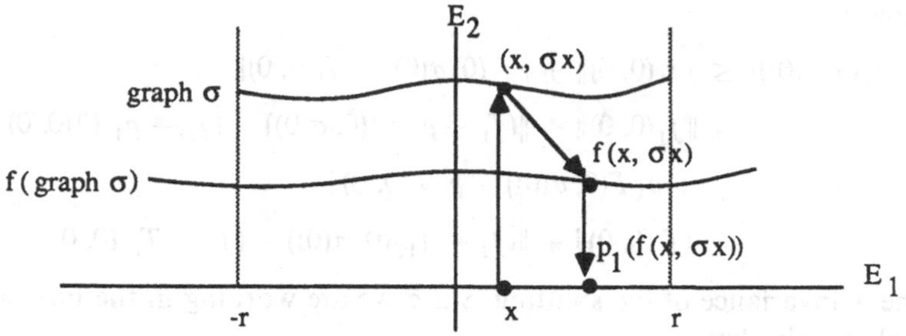

Figure 5.1.

than or equal to 1. First we show that Γ_f is well defined on $\text{Lip}_1(E_1(r), E_2(r))$. Next, we show that Γ_f is a contraction of $\text{Lip}_1(E_1(r), E_2(r))$ in the C^0 metric. Finally, we find g as the fixed point guaranteed by the contraction mapping principle.

Lemma 5.4. *If σ belongs to $\text{Lip}_1(E_1(r), E_2(r))$, we have the estimate*

$$\text{Lip}(f_1 \circ (\text{id}, \sigma) - T_1) \le \text{Lip}(f - T).$$

Proof. Since $f_1 \circ (\text{id}, \sigma) - T = p_1 \circ (f - T) \circ (\text{id}, \sigma)$, we have

$$\text{Lip}(f_1 \circ (\text{id}, \sigma) - T_1) \le \text{Lip}\, p_1\, \text{Lip}(f - T)\text{Lip}(\text{id}, \sigma) = \text{Lip}(f - T). \quad \square$$

Lemma 5.5. *If $\varepsilon > 0$ is less than $1/\lambda$ and $\text{Lip}(f - T)$ is less than ε, then for all σ in $\text{Lip}_1(E_1(r), E_2(r))$, the map $f_1 \circ (\text{id}, \sigma)$ is a homeomorphism. Furthermore, the inverse is a Lipschitz function whose Lipschitz constant satisfies*

$$\text{Lip}([f_1 \circ (\text{id}, \sigma)]^{-1}) \le \frac{1}{(1/\lambda - \varepsilon)}.$$

Proof. From above we have $\text{Lip}(f_1 \circ (\text{id}, \sigma) - T_1) \le \text{Lip}(f - T) \le \varepsilon$. Thus if ε is less than $1/\lambda$ which is less than $\|T_1^{-1}\|^{-1}$, the Lipschitz inverse function theorem applies. (See Appendix I of this chapter.)

Therefore $f_1 \circ (\text{id}, \sigma)$ is a homeomorphism and we have

$$\text{Lip}([f_1 \circ (\text{id}, \sigma)]^{-1}) \le \frac{1}{[\|T^{-1}\|^{-1} - \text{Lip}(f_1 \circ (\text{id}, \sigma) - T_1)]} \le \frac{1}{(1/\lambda - \varepsilon)}. \quad \square$$

Lemma 5.6. *Let $0 < 2\varepsilon < 1/\lambda - 1$. Suppose that $\text{Lip}(f - T) < \varepsilon$ and $\|f(0)\| < r(1/\lambda - 1 - 2\varepsilon)$; then for all σ in $\text{Lip}_1(E_1(r), E_2(r))$, $f_1 \circ (\text{id}, \sigma)$ overflows $E_1(r)$.*

Proof. By the proposition in Appendix I, the image $f_1 \circ (\text{id}, \sigma)(E_1(r))$ contains the ball of radius $r(1/\lambda - \varepsilon)$ about $f_1(0, \sigma(0))$ since $\text{Lip}([f_1 \circ (\text{id}, \sigma)]^{-1}) < 1/(1/\lambda - \varepsilon)$. Consequently, $f_1 \circ (\text{id}, \sigma)(E_1(r))$ also contains the ball of radius $\rho = r(1/\lambda - \varepsilon) - \|f_1(0, \sigma(0))\|$ about the origin.

However,

$$\begin{aligned}\|f_1(0, \sigma(0))\| &\leq \|f_1(0, 0)\| + \|f_1(0, \sigma(0)) - f_1(0, 0)\| \\ &\leq \|f_1(0, 0)\| + \|(f_1 - p_1 T)(0, \sigma(0)) - (f_1 - p_1 T)(0, 0) \\ &\quad + p_1 T(0, \sigma(0)) - p_1 T(0, 0)\| \\ &= \|f_1(0, 0)\| + \|(f_1 - T_1)(0, \sigma(0)) - (f_1 - T_1)(0, 0)\|\end{aligned}$$

by the T invariance of the splitting. Since we are working in the box norm, though, we also have

$$\begin{aligned}\|f_1(0, \sigma(0)) &\leq \|f(0, 0)\| + \|(f - T)(0, \sigma(0)) - (f - T)(0, 0)\| \\ &\leq \|f(0)\| + \varepsilon r.\end{aligned}$$

Putting this all together, we get

$$\rho \geq r\left(\frac{1}{\lambda} - 2\varepsilon\right) - \|f(0)\| > r. \qquad \square$$

Lemma 5.7. *Let* $0 < 2\varepsilon < 1 - \lambda$ *and* $\delta < r \min\{1/\lambda - 1 - 2\varepsilon, 1 - \varepsilon - \lambda\}$. *If* f *satisfies* $\mathrm{Lip}(f - T) < \varepsilon$ *and* $\|f(0)\| < \delta$, *then for all* σ *in* $\mathrm{Lip}_1(E_1(r), E_2(r))$ *the map* $\Gamma_f(\sigma)$ *is well defined on* $E_1(r)$ *and* $\Gamma_f(\sigma)$ *is in* $Lip_1(E_1(r), E_2(r))$.

PROOF. Notice that for $0 < \lambda < 1$, $1/\lambda - 1 > 1 - \lambda$ so the first statement follows directly from the preceding lemma and the definition of Γ_f. For the second, we estimate

$$\begin{aligned}\mathrm{Lip}\,\Gamma_f(\sigma) &\leq \mathrm{Lip}(f_2 \circ (\mathrm{id}, \sigma))\mathrm{Lip}([f_1 \circ (\mathrm{id}, \sigma)]^{-1}) \\ &\leq \mathrm{Lip}(f_2 \circ (\mathrm{id}, \sigma) \leq \mathrm{Lip}\, f_2\, \mathrm{Lip}(\mathrm{id}, \sigma) \\ &\leq \mathrm{Lip}\, f_2 \\ &\leq \mathrm{Lip}\, T_2 + \mathrm{Lip}(p_2 \circ (f - T)) \\ &\leq \lambda + \varepsilon \leq 1.\end{aligned}$$

Now we need only check $\Gamma_f(\sigma)(E_1(r)) \subset E_2(r)$. In fact, it is enough to check that $f_2 \circ (\mathrm{id}, \sigma)$ $(E_1(r)) \subset E_2(r)$ since $\Gamma_f(\sigma)(E_1(r)) = [f_2 \circ (\mathrm{id}, \sigma)] \circ [f_1 \circ (\mathrm{id}, \sigma)]^{-1}(E_1(r))$ and the homeomorphism $f_1 \circ (\mathrm{id}, \sigma)$ overflows $E_1(r)$ so $[f_1 \circ (\mathrm{id}, \sigma)]^{-1}(E_1(r)) \subset E_1(r)$. Now for x in $E_1(r)$

$$\begin{aligned}\|f_2(x, \sigma(x))\| &\leq \|f_2(x, \sigma(x)) - p_2 T(x, \sigma(x))\| + \|p_2 T(x, \sigma(x))\| \\ &\leq \|f_2(x, \sigma(x)) - p_2 T(x, \sigma(x))\| + \|T_2\|\, \|\sigma(x)\| \\ &\leq \|p_2 f(x, \sigma(x)) - p_2 T(x, \sigma(x))\| + \lambda r \\ &\leq \|(f - T)(x, \sigma(x))\| + \lambda r\end{aligned}$$

$$\leq \|(f - T)(x, \sigma(x)) - (f - T)(0, 0)\| + \|(f - T)(0, 0)\| + \lambda r$$
$$\leq \text{Lip}(f - T) \|(x, \sigma(x))\| + \|f(0)\| + \lambda r$$
$$\leq \varepsilon r + \delta + \lambda r \leq r. \qquad \square$$

From now on, to take advantage of the previous lemmas, we will assume that the following hold:

$$\text{Lip}(f - T) < \varepsilon < (1 - \lambda); \qquad \|f(0)\| < \delta < r \min\left\{\frac{1}{\lambda} - 1 - 2\varepsilon, 1 - \varepsilon - \lambda\right\}.$$

We now set out to show that Γ_f contracts $\text{Lip}_1(E_1(r), E_2(r))$ with respect to the sup norm; in fact we will show more, that f diminishes the vertical distance between any point and the graph of any witness σ in $\text{Lip}_1(E_1(r), E_2(r))$.

Lemma 5.8. *Let (x, y) be a point of $E_1(r) \times E_2(r)$ such that $f_1(x, y)$ lies in $E_1(r)$. For all σ in $\text{Lip}_1(E_1(r), E_2(r))$ the following bound holds:*

$$\|f_2(x, y) - \Gamma_f\sigma(f_1(x, y))\| \leq (\lambda + 2\varepsilon) \|y - \sigma(x)\|.$$

PROOF.

$$\begin{aligned}\|f_2(x, y) - \Gamma_f\sigma(f_1(x, y))\| &\leq \|f_2(x, y) - f_2(x, \sigma(x))\| \\ &\quad + \|f_2(x, \sigma(x)) - \Gamma_f\sigma f_1(x, y)\| \\ &= \|f_2(x, y) - f_2(x, \sigma(x))\| \\ &\quad + \|\Gamma_f\sigma(f_1(x, \sigma(x))) - \Gamma_f\sigma(f_1(x, y))\|,\end{aligned}$$

by definition of Γ_f. Thus $\|f_2(x, y) - \Gamma_f\sigma(f_1(x, y))\| \leq \text{Lip}\, f_2 \|(x, y) - (x, \sigma(x))\| + \text{Lip}\, \Gamma_f\sigma \|f_1(x, \sigma(x)) - f_1(x, y)\|$. But $\text{Lip}\, f_2 = \text{Lip}(f_2 - p_2 T +$

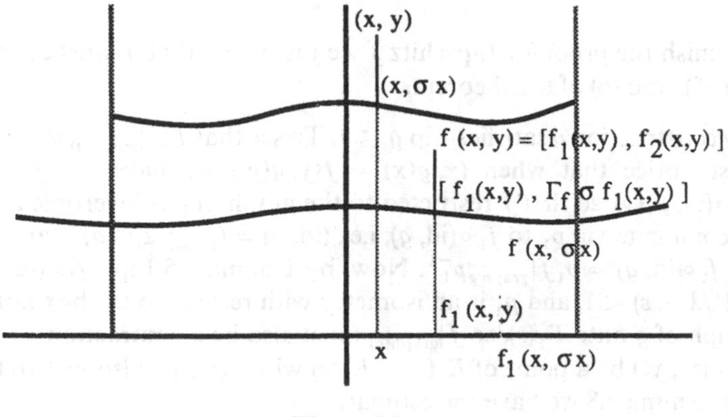

Figure 5.2.

$p_2 T) \leq \varepsilon + \lambda$, and by Lemma 5.7, Lip $\Gamma_f \sigma \leq 1$, so we have:

$$
\begin{aligned}
&\|f_2(x, y) - \Gamma_f\sigma(f_1(x, y))\| \\
&\quad \leq (\lambda + \varepsilon)\|y - \sigma(x)\| + \|f_1(x, \sigma(x)) - f_1(x, y)\| \\
&\quad \leq (\lambda + \varepsilon)\|y - \sigma(x)\| + \|(f_1 - p_1 T)(x, \sigma(x)) - (f_1 - p_1 T)(x, y)\| \\
&\qquad + \|p_1 T(x, \sigma(x)) - p_1 T(x, y)\| \\
&\quad \leq (\lambda + \varepsilon)\|y - \sigma(x)\| + \mathrm{Lip}(f_1 - p_1 T)\|y - \sigma(x)\| + 0 \\
&\quad \leq (\lambda + \varepsilon)\|y - \sigma(x)\| + \varepsilon\|y - \sigma(x)\|.
\end{aligned}
$$

□

Lemma 5.9. *In the above situation Γ_f contracts C^0 distances by a factor at most $\lambda + 2\varepsilon$.*

PROOF. Let σ_1 and σ_2 be two maps in $\mathrm{Lip}_1(E_1(r), E_2(r))$ and z be a point of $E_1(r)$. Applying Lemma 5.8 to $\sigma = \sigma_2$ at the point $(x, y) = ([f_1 \circ (\mathrm{id}, \sigma_1)]^{-1}(z)), \sigma_1([f_1 \circ (\mathrm{id}, \sigma_1)]^{-1}(z))$ we see that

$$
\begin{aligned}
\|\Gamma_f\sigma_1(z) - \Gamma_f\sigma_2(z)\| \leq (\lambda + 2\varepsilon)\|&\sigma_1[f_1 \circ (\mathrm{id}, \sigma_1)]^{-1}(z) \\
&- \sigma_2[f_1 \circ (\mathrm{id}, \sigma_1)]^{-1}(z)\|.
\end{aligned}
$$

Taking the supremum over all z then gives the result. □

Since $\mathrm{Lip}_1(E_1(r), E_2(r))$ is a closed subspace of the Banach space $C^0(E_1(r), E_2(r))$ and hence complete, the contraction mapping principle gives:

Proposition 5.10. *If* $\mathrm{Lip}(f - T) < \varepsilon < (1 - \lambda)/2$ *and*

$$\left\|f(0)\right\| \leq \delta < r \min\left\{\frac{1}{\lambda - 1 - 2\varepsilon}, 1 - \varepsilon - \lambda\right\}$$

then the graph transform Γ_f has a unique fixed point g in $\mathrm{Lip}_1(E_1(r), E_2(r))$.

□

To finish the proof for Lipschitz f we just verify that g satisfies conditions (1), (2), (5), and (6) of the theorem.

(1) By construction g satisfies Lip $g \leq 1$. To see that $f^{-1}|_{\mathrm{graph}(g)}$ is contraction, first notice that when $(x, g(x)) = f(y, g(y))$ we have $x = f_1(y, g(y)) = f_1 \circ (\mathrm{id}, g)(y)$, so that f restricted to the graph of g is invertible and in fact is conjugate via p_1 to $f_1 \circ (\mathrm{id}, g)$, i.e., $(\mathrm{id}, g) = (p_1|\mathrm{graph}\ g)^{-1}$ and $f_1 = p_1 f$ so $f_1 \circ (\mathrm{id}, g) = p_1 f|_{\mathrm{graph}\ g} p_1^{-1}$. Now, by Lemma 5.5 $\mathrm{Lip}([f_1 \circ (\mathrm{id}, g)]^{-1}) \leq 1/(1/\lambda - \varepsilon) < 1$, and p_1 is an isometry with respect to the box norm of the graph of g onto $E_1(r)$, so $f|_{\mathrm{graph}(g)}$ must also be a contraction.

(2) Let (x', y') be a point of $E_1(r) \times E_2(r)$ with $f(x', y')$ also in $E_1(r) \times E_2(r)$. By Lemma 5.8 we have the estimate

$$\|f_2(x', y') - g(f_1(x', y'))\| \leq (\lambda + 2\varepsilon)\|y' - g(x')\|.$$

Repeating the process, we see that if the first n iterates $(x', y'), f(x', y'), \ldots, f^n(x', y') = (x, y)$ lie in $E_1(r) \times E_2(r)$, we have

$$\begin{aligned}\|y - g(x)\| &= \|p_2 f^n(x', y') - g p_1 f^n(x', y')\| \\ &\le (\lambda + 2\varepsilon)^n \|y' - g(x')\| \le (\lambda + 2\varepsilon)^n 2r.\end{aligned}$$

Letting n go to infinity we see that if (x, y) is in $\bigcap_{n=0}^{\infty} f^n(E_1(r) \times E_2(r))$, y must equal $g(x)$, so $\bigcap_{n=0}^{\infty} f^n(E_1(r) \times E_2(r)) \subset \text{graph}(g)$.

On the other hand, since $\text{graph}(g) = f(\text{graph}(g)) \cap E_1(r) \times E_2(r)$ it follows that $\text{graph}(g) \subset \bigcap_{n=0}^{\infty} f^n(E_1(r) \times E_2(r))$ and we are done.

(5) and (6) If $f(0) = 0$ by (2) 0 is the only fixed point of f contained in the graph of g and all other points z of graph of g approach it under backwards iteration: $f^{-n}(z) \to 0$ as $n \to \infty$.

II. The Case of Smooth f

We begin the study of the delicate question of smoothness of the unstable manifold by showing that it is C^1 whenever f is C^1. The idea of the proof is this: if there is a C^1 function g whose graph is f invariant, then the derivative of f maps the tangent space of the graph to the tangent space of the graph, i.e.,

$$Df_{(x,g(x))}(T_{(x,g(x))} \text{ graph } g) = T_{f(x,g(x))} \text{ graph } g$$

or

$$Df_{(x,g(x))}(\text{graph } Dg_x) = \text{graph}(Dg_{f_1(x,g(x))})$$

we are lead to consider a new (global, linear) graph transform (Figure 5.3). The contraction mapping principle again will allow us to find a function $\sigma: E_1(r) \to L_1(E_1, E_2)$, where $L_1(E_1, E_2)$ is the space of continuous linear maps from E_1 to E_2 of norm less than or equal to 1, which satisfies

$$\Gamma_{Df}\sigma(x) = \sigma(f_1(x, g(x))).$$

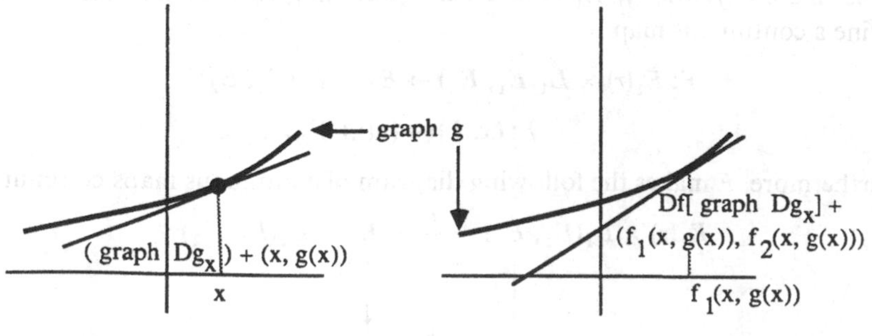

Figure 5.3.

Here and below, to simplify notation we write Df for $Df_{(x,g(x))}$. Finally, we show that σ is, in fact, the derivative of g.

First, let us examine some properties of the graph transform associated to a linear map S close to a hyperbolic map T with $\|T^{-1}|_{E_1}\| < \lambda$, $\|T|_{E_2}\| < \lambda$.

Lemma 5.11. *There is an $\varepsilon > 0$ such that when $\|S - T\| < \varepsilon$, the graph transform Γ_S is defined on $L_1(E_1, E_2)$ and sends it into itself.*

Moreover, Γ_S is Lipschitz on $L_1(E_1, E_2)$ with constant less than or equal to $(\lambda + 2\varepsilon)$.

PROOF. First note that since a linear map is Lipschitz with constant equal to its norm, $L_1(E_1, E_2) \subset \mathrm{Lip}_1\,(E_1(r), E_2(r))$ for all r. Thus choosing an ε in Lemma 5.7 independently of r we see that Γ_S is defined on $\mathrm{Lip}_1\,(E_1(r), E_2(r))$ for any r hence on $L_1(E_1, E_2)$.

Furthermore, since S is linear it sends linear subspaces to linear subspaces and Γ_S thus sends linear maps to maps with linear subspaces as graphs, that is, linear maps.

Finally, the Lipschitz constant of Γ_S is estimated by Lemma 5.9. □

Let us henceforth assume that ε is always sufficiently small so that the preceding lemma applies.

Lemma 5.12. *Let U_ε be an ε neighborhood of T in $L_1(E_1, E_2)$. The map $\Gamma\colon (S, K) \to \Gamma_S(K)$ is continuous from $U_\varepsilon \times L_1(E_1, E_2) \to L_1(E_1, E_2)$.*

PROOF. Let $S_i = p_i \circ S$. We can write $\Gamma_S(K) = S_2 \circ (\mathrm{id}, K) \circ [S_1 \circ (\mathrm{id}, K)]^{-1}$; and since inversion and composition are continuous on the space of linear maps Γ must also be continuous. □

Suppose that f is C^1 close to T on $E_1(r) \times E_2(r)$, so in particular, $\mathrm{Lip}(f - T) < \varepsilon$ and $\|Df(z) - T\| < \varepsilon$ for all z in $E_1(r) \times E_2(r)$. Let g be the map of $E_1(r)$ to $E_2(r)$ whose graph is the unstable manifold of f. We will now examine the graph of the derivative of g.

Setting $h = f_1 \circ (\mathrm{id}, g)\colon E_1(r) \to E_1$, the preceding two lemmas allow us to define a continuous map

$$F\colon E_1(r) \times L_1(E_1, E_2) \to E_1 \times L_1(E_1, E_2),$$

$$F\colon (x, L) \mapsto (h(x), \Gamma_{Df} L).$$

Furthermore, F makes the following diagram of continuous maps commute:

$$\begin{array}{ccc} E_1(r) \times L_1(E_1, E_2) & \xrightarrow{F} & E_1 \times L_1(E_1, E_2) \\ \downarrow & & \downarrow \\ E_1(r) & \xrightarrow{h} & E_1, \end{array} \qquad \text{(I)}$$

where the vertical maps are projection on the first factor. Lemma 5.11 and the definition of h give the following:

Lemma 5.13. $\|F(x, L) - F(x, K)\| \leq (\lambda + 2\varepsilon)\|L - K\|$, *uniformaly on* $E_1(r)$ *and, furthermore,* $E_1(r) \subset h(E_1(r))$, $\text{Lip}\, h^{-1} < 1$. □

Let $\Gamma^0(E_1(r), E_1(r) \times L_1(E_1, E_2))$ be the space of continuous sections of the trivial bundle $E_1(r) \times L_1(E_1, E_2) \to E_1(r)$, with the uniform metric; for sections σ_1 and σ_2:

$$d(\sigma_1, \sigma_2) = \sup_{x \in E_1(r)} \|\Pi_2 \sigma_1(x) - \Pi_2 \sigma_2(x)\|,$$

where Π_2 is the projection on the second factor of $E_1(r) \times L_1(E_1, E_2)$. Notice that this space of sections is isometric, via composition with Π_2, with the complete space of continuous maps of $E_1(r)$ to $L_1(E_1, E_2)$ and that the images of sections thereby correspond to graphs. Thus we define a new graph transform Γ_F to be the automorphism $\Gamma_F: \tau \mapsto F\tau h^{-1}$ of $\Gamma^0(E_1(r), E_1(r) \times L_1(E_1, E_2))$; that is, $\Gamma_F(\tau)$ is the section whose image is the intersection of F (image of τ) with $E_1(r) \times L_1(E_1, E_2)$.

Lemma 5.14. Γ_F *has a unique fixed point* σ *which satisfies*

$$\Gamma_{Df}(\Pi_2 \sigma(x)) = \Pi_2 \sigma h(x) = \Pi_2 \sigma f_1(x, g(x)).$$

PROOF. Lemma 5.13 says that Γ_F is a contraction of the complete space $\Gamma^0(E_1(r), E_1(r) \times L_1(E_1, E_2))$, so we are done by the contraction mapping principle. □

Definition 5.15. Let Y and Z be two metric space. Suppose that h_1 and h_2 are two functions from a neighborhood of x in Y to Z, with $h_1(x) = h_2(x)$. We say that h_1 and h_2 are *tangent at* x if and only if

$$\text{Lip}_x(h_1, h_2) = \limsup_{y \to x} (d(h_1(y), h_2(y)))/d(x, y) = 0.$$

That is, the Lipschitz distance from h_1 to h_2 at x is 0.

Examples. (1) If E_1 and E_2 are normed vector spaces and L_1 and L_2 are two continuous linear maps of E_1 to E_2, then, independently of x,

$$\text{Lip}_x(L_1, L_2) = \|L_1 - L_2\|.$$

(2) If f is a map from an open subset of a Banach space E_1 to a Banach space E_2 and L is a continuous linear map of E_1 to E_2, L is the derivative of f at x if and only if $f(x + y)$ and $f(x) + L(y)$ are tangent at $y = 0$, which is in this case means merely that

$$\lim_{y \to 0} \frac{\|f(x) - f(x + y) - L(y)\|}{\|y\|} = 0$$

Proposition 5.16. *When f is C^1, the fixed point g of Γ_f is C^1 with derivative $\Pi_2\sigma$, where σ is the fixed point of Γ_F.*

PROOF. Suppose we had the following estimate, where $h = f_1 \circ (\text{id}, g)$, and we are considering functions of y:

$$(*)\qquad \begin{aligned} &\text{Lip}_0[(\Gamma_f g)(h(x) + y), gh(x) + \Gamma_{Df}(\Pi_2\sigma(x))(y)] \\ &\qquad \leq (\lambda + 2\varepsilon)\text{Lip}_0[g(x + y), g(x) + \Pi_2\sigma(x)(y)]. \end{aligned}$$

That is applying Γ_F makes g and our candidate more tangent.

Rewriting $(*)$, when $h(x)$ is in $E_1(r)$, as

$$(**)\qquad \begin{aligned} &\text{Lip}_0[g(h(x) + y), gh(x) + \Pi_2\sigma(h(x))(y)] \\ &\qquad \leq (\lambda + 2\varepsilon)\text{Lip}_0[g(x + y), g(x) + \Pi_2\sigma(x)(y)] \end{aligned}$$

and repeating the estimation, we see that for x in $E_1(r)$

$$\begin{aligned} &\text{Lip}_0[g(h^{-n}(x) + y), g(h^{-n}(x)) + \Pi_2\sigma(h^{-n}(x))(y)] \\ &\qquad \geq \left(\frac{1}{\lambda + 2\varepsilon}\right)^n \text{Lip}_0[g(x + y), g(x) + \Pi_2\sigma(x)(y)]. \end{aligned}$$

Therefore if for some x in $E_1(r)$, $\text{Lip}_0[g(x + y), g(x) + \Pi_2\sigma(x)(y)] = \delta > 0$, then there is a sequence x_n in $E_1(r)$ such that

$$\text{Lip}_0[g(x_n + y), g(x_n) + \Pi_2\sigma(x_n)(y)] \to \infty.$$

But the Lipschitz constants of g and $\Pi_2\sigma(x_n)$ are both bounded by 1, so for all x_n in $E_1(r)$

$$\text{Lip}_0[g(x_n + y), g(x_n) + \Pi_2\sigma(x_n)(y)] \leq 2.$$

Consequently we must have

$$\text{Lip}_0[g(x + y), g(x) + \Pi_2\sigma(x)(y)] = 0$$

for all x in $E_1(r)$; therefore $Dg(x) = \Pi_2\sigma(x)$. Now on to establishing $(*)$. First we expand

$$\begin{aligned} &(\Gamma_f g)(h(x) + y) - gh(x) - \Gamma_{Df}[\Pi_2\sigma(x)](y) \\ &\qquad = (\Gamma_f g)(h(x) + y) - gh(x) - \Gamma_{Df}[g \circ (\text{id} + x) - g(x)](y) \\ &\qquad\quad + \Gamma_{Df}[g \circ (\text{id} + x) - g(x)](y) - \Gamma_{Df}[\Pi_2\sigma(x)](y). \end{aligned}$$

Thus, we can estimate

$$\begin{aligned} &\text{Lip}_0[(\Gamma_f g)(h(x) + y), gh(x) + \Gamma_{Df}[\Pi_2\sigma(x)](y)] \\ &\quad \leq \text{Lip}_0[(\Gamma_f g)(h(x) + y) - gh(x), \Gamma_{Df}[g \circ (\text{id} + x) - g(x)](y)] \\ &\qquad + \text{Lip}_0[\Gamma_{Df}[g \circ (\text{id} + x) - g(x)](y), \Gamma_{Df}[\Pi_2\sigma(x)](y)] = (1) + (2). \end{aligned}$$

First we attack (2). Let $k = p_1 Df(\mathrm{id}, g\circ(\mathrm{id}+x) - g(x))$. Since Df is close to T, k is surjective. Take w' such that $h(w') = y$. As k^{-1} is a contraction, $\|w'\| < \|y\|$. Applying Lemma 5.8 to $Df_{(x,g(x))}$ gives

$$\|\Gamma_{Df}[\Pi_2\sigma(x)](y) - \Gamma_{Df}[g\circ(x+\mathrm{id}) - g(x)](y)\|$$
$$\leq (\lambda + 2\varepsilon)\|[\Pi_2\sigma(x)](w') - g(x+w') + g(x)\|.$$

Therefore we get

$$\frac{\|\Gamma_{Df}[\Pi_2\sigma(x)](y) - \Gamma_{Df}[g\circ(x+\mathrm{id}) - g(x)](y)\|}{\|y\|}$$
$$\leq (\lambda + 2\varepsilon)\frac{\|g(x+w') - g(x) - \Pi_2\sigma(x)w'\|}{\|w'\|},$$

that is $(2) \leq (\lambda + 2\varepsilon)\mathrm{Lip}_0[\Pi_2\sigma(x), g\circ(x+\mathrm{id}) - g(x)]$. To finish with $(*)$ we must show $(1) = 0$. Choose a w such that $h(x+w) = h(x) + y$; since $\mathrm{Lip}(h^{-1}) < 1$, $\|w\| \leq \|y\|$. Now we can rewrite

$$(\Gamma_f g)(h(x) + y) = f_2(x+w, g(x+w)), \qquad gh(x) = f_2(x, g(x))$$

and, by the choice of w',

$$\Gamma_{Df}[g\circ(x+\mathrm{id}) - g(x)](y) = p_2 Df[w', g(x+w') - g(x)].$$

Thus we can express

$$\begin{aligned}(3) &= \|(\Gamma_f g)(h(x)+y) - gh(x) - \Gamma_{Df}[g\circ(x+\mathrm{id}) - g(x)](y)\|\\ &= \|p_2 f(x+w, g(x+w)) - p_2 f(x, g(x)) - p_2 Df[w', g(x+w') - g(x)]\|\\ &= \|p_2 Df[w, g(x+w) - g(x)] + p_2 R[w, g(x+w) - g(x)]\\ &\quad - p_2 Df[w', g(x+w') - g(x)]\|\end{aligned}$$

by Taylor's theorem, with $\lim_{w\to 0}\|R(w, g(x+w) - g(x))\|/\|w\| = 0$. Simplifying, we get

$$(3) = \|p_2 Df(w - w', g(x+w) - g(x+w')) + p_2 R[w, g(x+w) - g(x)]\|.$$

Since $\mathrm{Lip}(g) \leq 1$ and Df is linear we can estimate

$$(4) = \frac{(3)}{\|y\|} = \frac{\|(\Gamma_f g)(h(x)+y) - gh(x) - \Gamma_{Df}[g\circ(x+\mathrm{id}) - g(x)](y)\|}{\|y\|}$$
$$\leq \|p_2 Df\|\frac{\|w - w'\|}{\|y\|} + \frac{\|R(w, g(x+w) - g(x))\|}{\|y\|}.$$

Note that (1) is just the limit of (4) as y approaches 0 and, since $\|w\| \leq \|y\|$, so $\lim_{\|y\|\to 0}\|R\|/\|y\| = \lim_{\|w\|\to 0}\|R\|/\|w\| = 0$, our latest estimate shows that $(1) = 0$, provided $\|w - w'\|/\|y\|$ goes to 0 with $\|y\|$.

Recall the definition of w and w':

$$p_1 Df(w', g(x + w') - g(x)) = y,$$
$$h(x + w) - h(x) = y.$$

Taylor's theorem now tells us

$$p_1 f(x + w', g(x + w')) - p_1 f(x, g(x))$$
$$= p_1 Df(w', g(x + w') - g(x)) + R(w', g(x + w') - g(x)).$$

But the first expression is just $h(x + w') - h(x)$, and, since the second expression is $y - R$, we can use the triangle inequality and the fact that $\mathrm{Lip}(h^{-1}) < 1$ to estimate $\|w - w'\| \leq \|R(w', g(x + w') - g(x))\|$. Finally, we see $\lim_{\|y\| \to 0}(\|w - w'\|/\|y\|) \leq \lim_{\|y\| \to 0}(\|R\|/\|y\|) = 0$, and Lemma 5.16 is proven. □

If we check condition (4) we will have finished proving the unstable manifold theorem for $C^1 f$. Suppose $f(0) = 0$, $Df_0 = T$ and notice that σ satisfies

$$\Gamma_{Df}[\Pi_2(\sigma(x))] = \Pi_2 \sigma f_1(x, g(x)).$$

So $Df_{(0, g(0))} = T$,

$$\Gamma_T[\Pi_2 \sigma(0)] = \Pi_2 \sigma(0).$$

Now since E_1 is the unique T invariant graph, the arguments in the proof of Lemma 5.16 show that the graph of g is tangent to E_1 at 0.

Luckily, increasing the smoothness from 1 to r is easier than going from Lipschitz to C^1. Referring to diagram (I) we need only show that the invariant section σ is C^{r-1}. This is just a special case of the C^r Section Theorem 5.18 below.

Definition 5.17. Let $\Pi: E \to X$ be a vector bundle with a metric space base. We say a metric d on E is *admissible* when:

(1) it induces a norm on each fiber;
(2) there is a complementary bundle E' over X and an isomorphism of $E \oplus E'$ and $X \times A$, where A is a Banach space and the product metric on $X \times A$ induces d on E;
(3) the projection of $X \times A$ onto E is of norm 1.

Recall that a map φ between two metric spaces Y_1 and Y_2 is said to α-Hölder, $0 < \alpha \leq 1$, if there is a constant K such that for all x and y in Y_1

$$d(\varphi(x), \varphi(y)) \leq K(d(x, y))^\alpha.$$

Theorem 5.18 (C^r Section Theorem). *Let $\Pi: E \to X$ be a vector bundle over the metric space X, with an admissible metric on E. Let X_0 be a subset of X and D*

be the disc bundle of radius C in E, where $C > 0$ is a finite constant. Let D_0 be the restriction of D to X_0; $D_0 = D \cap \Pi^{-1}(X_0)$.

Let h be an overflowing homeomorphism of X_0 into X, that is, $X_0 \subset h(X_0)$. Let $F: D_0 \to D$ be a map which covers h. Suppose that there is a constant k, $0 \leq k < 1$, such that, for all x in X_0, the restriction of F to the fiber over x, $F_x: D_x \to D_{h(x)}$, is Lipschitz with constant at most k.

Then:

(a) *There is a unique section $\sigma: X_0 \to D_0$ such that F (image of σ) $\cap D_0$ = image of σ.*
(b) *If F is continuous then so is σ.*
(c) *If, moreover, h^{-1} is Lipschitz with $\mathrm{Lip}(h^{-1}) = \mu$, F is α-Hölder, and $k\mu^\alpha < 1$, then σ is α-Hölder. In particular, when $\alpha = 1$, σ is Lipschitz.*
(d) *If, moreover, X, X_0, and E are C^r manifolds ($r \geq 1$), h and F are C^r, $F^{(j)}$ and $(h^{-1})^{(j)}$ are bounded for $1 \leq j \leq r$, Lipschitz for $1 \leq j < r$, and $k\mu^r < 1$, where $\mu = \mathrm{Lip}(h^{-1})$, then σ is C^r.*

REMARK. On a connected manifold of finite diameter, a bounded rth derivative guarantees a bounded and Lipschitz $(r-1)$th derivative.

Theorem 5.18 allows us to prove the unstable manifold theorem by induction as follows. We already know g is C^1 whenever f is; suppose we know that g is C^{s-1} whenever f is. With F and h as in diagram (I) we have $\mathrm{Lip}(h^{-1}) < 1$ and $\mathrm{Lip}(F_x) \leq (\lambda + 2\varepsilon) < 1$ so that the hypothesis of 5.18(d) hold for $r = s - 1$. Thus we can find a C^{s-1} invariant section σ which gives the derivative of g and hence g is C^s and we have finally finished proving all the conclusions of the unstable manifold theorem. □

This same reasoning will allow us to inductively demonstrate 5.18(d) itself.

PROOF OF THEOREM 5.18. We may assume, without loss of generality, that E is trivial. If it is not, the existence of admissible metric allows us to replace E by $E \oplus E'$ and F by the composition

$$F': E \oplus E' \xrightarrow{p} E \xrightarrow{F} E \xrightarrow{i} E \oplus E',$$

where p is projection and i is inclusion. F' clearly covers h and has the same Lipschitz constant on any fiber; moreover, since the image of F' is contained in E so is the image of any section invariant under $\Gamma_{F'}$.

Let, then, $E = X \times A$ so we can write $\sigma(x) = (x, \sigma_2(x))$ in coordinates.

(a) The space of local sections of E, $\Gamma(X_0, D_0)$ is complete in the uniform topology and as usual the map $\Gamma_F: \sigma \to F \circ \sigma \circ h^{-1}$ is a contraction so it has a unique fixed point.
(b) If F is continuous Γ_F maps the space of continuous sections of D_0, $\Gamma^0(X_0, D_0)$, into itself, so the invariant section must also be continuous.
(c) Let H be the Hölder constant of F, that is $\forall e_1, e_2 \in E$, $d(F(e_1), F(e_2)) \leq H(d(e_1, e_2))^\alpha$.

Let σ be the invariant section from (b), and let $F_2 = \Pi_2 \circ F$, where Π_2 is the projection of $X \times A$ onto A. The vertical component σ_2 must then satisfy

$$\begin{aligned} d[\sigma_2(x), \sigma_2(y)] &= d[F_2(h^{-1}(x), \sigma_2 h^{-1}(x)), F_2(h^{-1}(y), \sigma_2 h^{-1}(y))] \\ &\leq d[F_2(h^{-1}(x), \sigma_2 h^{-1}(x)), F_2(h^{-1}(x), \sigma_2 h^{-1}(y))] \\ &\quad + d[F(h^{-1}(x), \sigma_2 h^{-1}(y)), F(h^{-1}(y), \sigma_2 h^{-1}(y))] \\ &\leq kd[\sigma_2 h^{-1}(x), \sigma_2 h^{-1}(y)] + Hd[h^{-1}(x), h^{-1}(y)]^\alpha \\ &\leq kd[\sigma_2 h^{-1}(x), \sigma_2 h^{-1}(y)] + H(\mu d(x, y))^\alpha. \end{aligned}$$

In fact, we may show by induction that

$$(***) \quad d[\sigma_2(x), \sigma_2(y)] \leq k^n d[\sigma_2 h^{-n}(x), \sigma_2 h^{-n}(y)] + H \sum_{j=1}^{n} (\mu^\alpha)^j k^{j-1} (d(x, y))^\alpha$$

holds for all n. We have just shown this holds for $n = 1$. Hence we can estimate

$$\begin{aligned} k^m d[\sigma_2 h^{-m}(x), \sigma_2 h^{-m}(y)] &\leq k^m \{kd[\sigma_2 h^{-m-1}(x), \sigma_2 h^{-m-1}(y)] \\ &\quad + H\mu^\alpha d[h^{-m}(x), h^{-m}(y)]^\alpha\} \\ &\leq k^{m+1} d[\sigma_2 h^{-m-1}(x), \sigma_2 h^{-m-1}(y)] \\ &\quad + Hk^m \mu^\alpha (\mu^m d(x, y))^\alpha \\ &\leq k^{m+1} d[\sigma_2 h^{-m-1}(x), \sigma_2 h^{-m-1}(y)] \\ &\quad + Hk^m (\mu^\alpha)^{m+1} d(x, y)^\alpha. \end{aligned}$$

If (***) holds for $n = m$

$$\begin{aligned} d[\sigma_2(x), \sigma_2(y)] &\leq k^m d[\sigma_2 h^{-m}(x), \sigma_2 h^{-m}(y)] + H \sum_{j=1}^{m} ((\mu^\alpha)^j k^{j-1} (d(x, y))^\alpha \\ &\leq k^{m+1} d[\sigma_2 h^{-m-1}(x), \sigma_2 h^{-m-1}(y)] + Hk^m (\mu^\alpha)^{m+1} (d(x, y))^\alpha \\ &\quad + H \sum_{j=1}^{m} (\mu^\alpha)^j k^{j-1} (d(x, y))^\alpha \end{aligned}$$

and thus (***) holds for $n = m + 1$. Therefore we must have $d[\sigma_2(x), \sigma_2(y)] \leq \lim_{n \to \infty} \{k^n d[\sigma_2 h^{-n}(x), \sigma_2 h^{-n}(y)] + H\sum_{j=1}^{\infty} (\mu^\alpha)^j k^{j-1} (d(x, y))^\alpha\}$. In particular since k^n goes to 0 and $d[\sigma_2 h^{-n}(x), \sigma_2 h^{-n}(y)] \leq 2C$, $d[\sigma_2(x), \sigma_2(y)] \leq [H(\mu^\alpha/(1 - k\mu^\alpha))](d(x, y))^\alpha$, that is, σ_2 is α-Hölder. Since $\sigma = (\mathrm{id}, \sigma_2)$, σ is also α-Hölder.

(d) We will mimic the end of our proof of the unstable manifold theorem. We know that the invariant section is Lipschitz. Regarding the tangent bundle to its graph as the invariant section of an auxilliary graph transform now allows us to inductively show that its tangent bundle is C^{r-1}.

Let $\bar{E} \to X$ be the vector bundle with fiber $L(T_x X, A)$ at x, where $L(T_x X, A)$ is the space of continuous linear maps of the tangent space at x of X to A.

This is where the derivative of σ_2 should be. Let $\bar{D}$ be the disc bundle of radius C associated to $\bar{E}$; its fiber at x is the subset of maps in $L(T_xX, A)$ with norm not greater than C. First let us show that $\bar{E}$ has a trivializing complement E' as in the definition of an admissible metric. We can always find a V such that $TX \oplus V$ is trivial. Put a product metric on $TX \oplus V$ and let $\bar{E}'$ have fiber $L(V_x, A)$ at x. $\bar{E} \oplus \bar{E}'$ is isomorphic to the trivial bundle with fiber $L(T_xX \oplus V_x, A)$ at x, with the operator norm in the fiber and product metric. If $L \in L_x(T_xX, A)$ then the operator norms of L and $L \oplus 0$ are equal, and E is admissible.

Next we define a map $\Gamma_{DF}\colon \bar{D}_0 \to \bar{D}$ which covers the homeomorphism $h\colon X_0 \to X$. For a linear map L of T_xX to A, $\Gamma_{DF}(L)$ will be the linear map of $T_{h(x)}X$ to A defined by

$$\Gamma_{DF}(L) = (\Pi_2 DF_{x,\sigma(x)}) \circ (\mathrm{Id}, L) \circ (Dh^{-1}_{h(x)}),$$

where Π_2 is the projection of $T_{h(x)}X \times A$ onto A and $DF_{(x,y)}$ is the derivative of F at (x, y). Note that $\Gamma_{DF}(L) = \Pi_2 D_1 F_{(x,\sigma(x))}(Dh^{-1})_{h(x)} + \Pi_2 D_2 F_{(x,\sigma(x))} L (Dh^{-1})_{h(x)}$, where $D_1 F_{(x,y)}$ is the derivative with respect to the X variables and $D_2 F_{(x,y)}$ is the derivative with respect to the A variables. The following diagram commutes:

$$\begin{array}{ccc} \bar{E}_0 & \xrightarrow{\Gamma_{DF}} & \bar{E} \\ \downarrow & & \downarrow \\ X_0 & \xrightarrow{h} & X. \end{array}$$

As before, if σ is differentiable, the derivative of σ_2 is a fixed point of Γ_{DF}. Let us calculate the contraction constant for Γ_{DF} restricted to a fiber. The assumption on the Lipschitz constants of F and h^{-1} give the following:

$$\|\Pi_2 D_2 F_{(x,\sigma(x))}\| \le k \quad \text{and} \quad \|Dh^{-1}\| < \mu.$$

Therefore

$$\|\Pi_2 D_2 F_{(x,\sigma(x))} \circ (L_1 - L_2) \circ Dh^{-1}\| \le \mu k \|L_1 - L_2\|,$$

and we see that Γ_{DF} is thus well defined, continuous, affine on fibers, with bounded translation and with a fiber contraction factor

$$\mathrm{Lip}((\Gamma_{DF})_x) \le \mu k < 1.$$

Therefore, we can find a finite C such that Γ_{DF} maps $\bar{D}_0$ to $\bar{D}$. Now we are almost ready to apply the induction hypothesis. Suppose F is C^2. Suppose that we knew that the invariant section for Γ_{DF} were the derivative of σ_2; then the fact that σ is C^1 would imply that Γ_{DF} was also C^1. In our case, $\mu(\mu k) < 1$, by applying the theorem in the C^1 case, we deduce that the invariant section of Γ_{DF} is C^1 so σ is C^2. We finish the induction as follows: suppose that we have proved the theorem when F and h are C^{r-1}, and k and μ satisfy $k\mu^{r-1} < 1$, then F has a C^{r-1} invariant section σ. The fiber contraction factor of Γ_{DF} is

(μk); since $\mu^{r-1}(\mu k)$ is strictly less than 1 and Γ_{DF} is C^{r-1}, Γ_{DF} admits a C^{r-1} invariant section, so the derivative of σ_2 is C^{r-1} and σ_2 is C^r.

We need only show, then, that the invariant section τ of Γ_{DF} which we know to be continuous, is in fact the derivative of σ_2. Working in charts about $h(x)$ and x, and mimicking the estimates used to demonstrate the C^1 stable manifold theorem, we see, fixing x and letting y alone vary,

$$\begin{aligned}\operatorname{Lip}_0[\sigma_2(h(x)+y), \sigma_2(h(x)) + \tau(h(x))(y)]\\ \le \mu k \operatorname{Lip}_0[\sigma_2(x+y), \sigma_2(x) + \tau(x)(y)].\end{aligned}$$

On the other hand, since σ and $\tau(x)$ are uniformly Lipschitz, $\operatorname{Lip}_0(\sigma_2(x+y), \sigma_2(x)(y))$ is bounded. Therefore $\operatorname{Lip}_0(\sigma_2(x+y), \sigma_2(x) + \tau(x)(y)) = 0$, so $\tau(x)$ is the derivative of σ at x. Thus, we are done with the unstable manifold and C^r section theorems. □

Corollary 5.19. *Suppose f is a C^2 diffeomorphism of a C^2 manifold M, with a closed hyperbolic invariant set Λ.*

Let $TM|_\Lambda = E^s \oplus E^u$ be a hyperbolic splitting for Tf. The bundles E^s and E^u are then Hölder, that is their transition functions may be taken to be Hölder.

PROOF. Choose a C^1 adapted metric on M so that

$$\|Tf|_{E^s}\| < \lambda < 1 \quad \text{and} \quad \|Tf^{-1}|_{E^u}\| < \lambda < 1.$$

Let μ be the Lipschitz constant of f^{-1} restricted to Λ, which exists since f is C^1. Let F^s and F^u be C^1 approximations to E^s and E^u which are close enough so that the derivative of f at x, $Df_x\colon F^s_x \oplus F^u_x \to F^s_{f(x)} \oplus F^u_{f(x)}$, can be written as a block matrix $\begin{pmatrix} A_x & B_x \\ C_x & D_x \end{pmatrix}$, with A_x, D_x nonsingular, satisfying $\|A_x\|, \|D_x^{-1}\| < \lambda < 1$; $\|B_x\|, \|C_x\| < \varepsilon$, uniformly in x.

Let E be the C^1 bundle whose fiber at x is $L(F^u_x, F^s_x)$; if ε is small enough, we can define a graph transform on the unit disc subbundle D of E over Λ:

$$\begin{aligned}&F(x,\sigma) = (f(x), \Gamma_{DF}\sigma),\\ &\text{where} \quad \Gamma_{DF}\sigma(x) = [B_x + A_x\sigma(x)] \circ [C_x + D_x\sigma(x)]^{-1}.\end{aligned}$$

Now the map with matrix $\begin{pmatrix} A_x & B_x \\ C_x & D_x \end{pmatrix}$ is close to the hyperbolic map $\begin{pmatrix} A_x & O \\ O & D_x \end{pmatrix}$ and, further, there is an α, $0 < \alpha < 1$, such that $(\lambda + 2\varepsilon)\mu^\alpha < 1$. We know, then, that F contracts fibers of the disc bundle D by a factor $(\lambda + 2\varepsilon) < 1$ (cf. Lemma 5.13). On the other hand, F is Lipschitz, in fact, C^1 since f is C^2. Since $(\lambda + 2\varepsilon)\mu^\alpha < 1$, the invariant section is α-Hölder. This section is the unstable bundle E^u. As usual by considering f^{-1} instead of f we get the result for E^s. □

We conclude with a special case of Theorem 5.2.

Theorem 5.20. *If f is a C^r diffeomorphism and p is a hyperbolic fixed point of f, then there are local stable and unstable manifolds $W^s_{loc}(p)$ and $W^u_{loc}(p)$. These manifolds are tangent at p to E^s and E^u, respectively, and are as smooth as f.* □

The manifolds $W^u(p) = \bigcup_{n \geq 0} f^n(W^u_{loc}(p))$ and $W^s(p) = \bigcup_{n \leq 0} f^n(W^s_{loc}(p))$ are C^r and are called the global unstable and stable manifolds of p. They are injectively immersed discs. If p is periodic with period n we define $W^u_f(p)$ analogously as the global unstable manifold of p with respect to f^n, and similarly for W^s.

Exercise 5.1. Some of the global hypothesis of Theorem 5.18 may be replaced by pointwise conditions. In particular, we leave it to the reader to attempt to reprove 5.18(d) under the weaker hypothesis that

$$(\mathrm{Lip}_{k(x)} h^{-1}))^r k_x \leq C < 1$$

uniformly in x, where $k_x = \mathrm{Lip}(F_x)$.

Appendix I

The Local Lipschitz Inverse Function Theorem

We begin with a lemma.

Lemma I.1. *Let F be a Banach space, X a metric space and f, g two continuous maps of X to F. Suppose that f is injective and f^{-1} is Lipschitz. If g satisfies* $\mathrm{Lip}(f - g) < [\mathrm{Lip}(f^{-1})]^{-1}$, *then g is also injective and*

$$\begin{aligned}\mathrm{Lip}(g^{-1}) &\leq ([\mathrm{Lip}(f^{-1})]^{-1} - \mathrm{Lip}(f - g))^{-1} \\ &= \mathrm{Lip}(f^{-1})/(1 - \mathrm{Lip}(g - f)\mathrm{Lip}(f^{-1})).\end{aligned}$$

Proof. It is enough to calculate that

$$\begin{aligned}\|g(x) - g(y)\| &\geq \|f(x) - f(y)\| - \|(g - f)(x) - (g - f)(y)\| \\ &\geq [[\mathrm{Lip}(f^{-1})]^{-1} - \mathrm{Lip}(f - g)]d(x, y).\end{aligned}$$

□

Theorem I.2. *Let f be a homeomorphism from an open subset of U of a Banach space E to an open subset V of a Banach space F, whose inverse is Lipschitz. Let h be a continuous, Lipschitz map from U to F satisfying* $\mathrm{Lip}(h)\mathrm{Lip}(f^{-1}) < 1$. *Let $g = f + h$. Then g is a homeomorphism of U onto an open subset of F, with Lipschitz inverse.*

Proof. To deduce the theorem from the lemma we must first show that g is open, or, what is the same, since f^{-1} is a homeomorphism, that $gf^{-1} = (f + h)f^{-1} = \mathrm{id} + hf^{-1}$ is open.

Let $v = hf^{-1}$. We can estimate

$$\lambda = \text{Lip } v \le \text{Lip } h \text{ Lip } f^{-1} < 1.$$

For x in V we show that, if the closed ball of radius r about x, $\bar{B}_r(x)$ is contained in V, then

$$\bar{B}_{(1-\lambda)r}[(\text{id} + v)(x)] \subset (\text{id} + v)\bar{B}_r(x)$$

so id + v is open.

After perhaps composing v with translations, we can take $x = 0$ and $v(x) = 0$. Let $s = (1 - \lambda)r$. We seek a local inverse to id + v, or, what amounts to the same thing, a map $w: \bar{B}_s(0) \to F$ which satisfies

$$(\text{id} + w)(\bar{B}_s(0)) \subset \bar{B}_r(0) \quad \text{and} \quad (\text{id} + v)(\text{id} + w) = \text{id}.$$

The last equation is equivalent to $w = -v(\text{id} + w)$, which leads us to look for w as a fixed point. Let $Z = \{w \in C^0(\bar{B}_s(0), F) | w(0) = 0 \text{ and } \text{Lip}(w) \le \lambda/(1 - \lambda)\}$. Z is complete in the uniform metric.

Furthermore, simple calculation shows that for w in Z, $(\text{id} + w)(\bar{B}_s(0)) \subset \bar{B}_r(0)$ and the map $-v(\text{id} + w)$ is also in Z. If we can show that $\Phi: Z \to Z$ defined by $\Phi(w) = -v(\text{id} + w)$ is a contraction, then its fixed point w is the function we seek. We calculate

$$\| -v(\text{id} + w) + v(\text{id} + w')\| \le \lambda \|(\text{id} + w) - (\text{id} + w') = \lambda \|w - w'\|$$

and noting $\lambda < 1$ we are done. □

Proposition I.3. *Let U be an open subset of a Banach space E and g be a homeomorphism of U onto an open subset of the Banach space F. If g^{-1} is Lipschitz with* $\text{Lip}(g^{-1}) < \lambda$, *then* $\bar{B}_{r/\lambda}(g(x)) \subset g(\bar{B}_r(x))$.

PROOF. We can, without loss of generality, again assume $x = g(x) = 0$. Let $v \ne 0$ be a point in $B_{r/\lambda}(0)$, and set $t_\infty = \sup\{t \ge 0 | [0, t]v \subset g[\bar{B}_r(0)]\}$. Since $g[B_r(0)]$ contains a neighborhood of 0, t_∞ is strictly positive. Moreover, by definition we have

$$(0, t_\infty)v \subset g(\bar{B}_r(0)).$$

On the other hand, we can estimate

$$\|g^{-1}(tv) - g^{-1}(t'v)\| \le \lambda |t - t'| \, \|v\|.$$

This implies that $\lim_{t \to t_\infty} g^{-1}(tv)$ exists and is in the closed ball $\bar{B}_r(0)$; therefore $t_\infty v$ belongs to $g(\bar{B}_r(0))$. To show that v belongs to $g(\bar{B}_r(0))$ we need only show that t_∞ is no less than 1. If t_∞ were less than 1, we would have

$$\|g^{-1}(t_\infty v)\| = \|g^{-1}(t_\infty v) - g^{-1}(0)\| \le \lambda t_\infty \|v\| < \lambda \|v\| \le r$$

and $t_\infty v$ would thus belong to $g(B_r(0))$. Since g is open, we could find an $\varepsilon > 0$ so that $[t_\infty, t_\infty + \varepsilon]v$ would also be in the open set $g(B_r(0))$ contradicting the definition of t_∞. □

We will also have occasion to use the better known C^r local inverse function theorem in what follows.

Appendix II

Irwin's Proof of the Stable Manifold Theorem

Let $E = E_1 \oplus E_2$ be a Banach space with an automorphism T which preserves the splitting. Suppose that T is hyperbolic, contracting E_1 and expanding E_2. In other words, for some λ, $0 < \lambda < 1$ we have

$$\|T_1\| < \lambda, \qquad \|T_2^{-1}\| < \lambda, \qquad \text{where} \quad T_i = T|_{E_i}.$$

Notice that in Chapter 5 we had T_1 expanding and T_2 contracting. The reason for the switch is that this time we will first find the stable manifold and then from this deduce the existence of the unstable manifold.

Once again, we furnish E with the box norm and retain the conventions of Chapter 5 about the projections p_i, $E_i(r)$, f_i, etc.

Lemma II.1. *If $f\colon E_1(r) \times E_2(r) \to E$ is a Lipschitz perturbation of T, with $\operatorname{Lip}(f - T) \le \varepsilon < 1 - \lambda$, then f preserves the family of cones parallel to $E_2(r)$. More precisely, if $x = (x_1, x_2)$ and $y = (y_1, y_2)$ are points in $E_1(r) \times E_2(r)$ with*

$$\|x_1 - y_1\| < \|x_2 - y_2\|$$

then

$$\begin{aligned}\|f_1(x) - f_1(y)\| &\le (\lambda + \varepsilon)\|x_2 - y_2\| \\ &< (\lambda^{-1} - \varepsilon)\|x_2 - y_2\| \le \|f_2(x) - f_2(y)\|.\end{aligned}$$

PROOF. On the one hand,

$$\begin{aligned}\|f_1(x) - f_1(y)\| &\le \|p_1(f - T)(x) - p_1(f - T)(y)\| + \|T_1 x_1 - T_1 y_1\| \\ &\le \varepsilon\|x - y\| + \lambda\|x_1 - y_1\| \\ &\le (\lambda + \varepsilon)\|x_2 - y_2\|.\end{aligned}$$

On the other hand,

$$\begin{aligned}\|f_2(x) - f_2(y)\| &\ge \|T_2 x_2 - T_2 y_2\| - \|p_2(f - T)x - p_2(f - T)y\| \\ &\ge \lambda^{-1}\|x_2 - y_2\| - \varepsilon\|x - y\| \\ &= (\lambda^{-1} - \varepsilon)\|x_2 - y_2\|.\end{aligned}$$

We are done, then, since $\lambda + \varepsilon < 1 < \lambda^{-1} - \varepsilon$. □

Definition II.2. For $f: E_1(r) \times E_2(r) \to E$ and $r \geq r' > 0$ we define $W_{r'}^s(f)$ to be

$$W_{r'}^s(f) = \{x \in E_1(r') \times E_2(r') | \forall n \geq 0, f^n(x) \text{ is defined and } \|f^n(x)\| \leq r\}.$$

Corollary II.3. *For $f: E_1(r) \times E_2(r) < \infty$ with $\text{Lip}(f - T) < \varepsilon < 1 - \lambda$, the set $W_r^s(f)$ is the graph of a function $g: A \to E_2(r)$, where A is the projection of $W_r^s(f)$ onto $E_1(r)$. Furthermore, g is Lipschitz with $\text{Lip}\, g \leq 1$ and $f | W_r^s(f)$ contracts distances. Therefore, f has at most one fixed point which, if it exists, attracts all other points of $W_r^s(f)$.*

Proof. Let $x = (x_1, x_2)$ and $y = (y_1, y_2)$ be any two points in $W_r^s(f)$. Our assertions about g are equivalent to showing that $\|x_2 - y_2\| \leq \|x_1 - y_1\|$.

Suppose that we had $\|x_2 - y_2\| > \|x_1 - y_1\|$, then the preceding lemma would tell us that

$$\|f_2(x) - f_2(y)\| \geq (\lambda^{-1} - \varepsilon)\|x_2 - y_2\|$$

and, by induction, for all $n \geq 1$

$$\|f_2^n(x) - f_2^n(y)\| \geq (\lambda^{-1} - \varepsilon)^n \|x_2 - y_2\|.$$

Since $\|f_2^n(x) - f_2^n(y)\| \leq 2r$ and $(\lambda^{-1} - \varepsilon)^n \to +\infty$ as $n \to \infty$ (recall $\lambda^{-1} - \varepsilon > 1$), we would have to conclude that $\|x_2 - y_2\| = 0$, in contradiction to our assumption.

Now, we must show that $f|_{W_r^s(f)}$ is a contraction. From the above, we know that for x, y in $W_r^s(f)$ we have $\|x - y\| = \|x_1 - y_1\|$. Moreover, since $f(W_r^s(f)) \subset W_r^s(f)$ we also have $\|f(x) - f(y)\| = \|f_1(x) - f_1(y)\|$. Therefore,

$$\begin{aligned}\|f(x) - f(y)\| &= \|f_1(x) - f_1(y)\| \\ &\leq \|(f_1 - T_1)(x) - (f_1 - T_1)(y)\| + \|T_1 x_1 - T_1 y_1\| \\ &\leq \varepsilon\|x - y\| + \lambda\|x_1 - y_1\| = (\lambda + \varepsilon)\|x - y\|.\end{aligned}$$

Since $(\lambda + \varepsilon) < 1$, f restricted to $W_r^s(f)$ is a contraction. □

Theorem II.4 (The Stable Manifold Theorem). *Let $T: E \to E$ be a hyperbolic automorphism as above. There is an $\varepsilon > 0$, depending only on λ, and, for all $r > 0$, a $\delta > 0$ such that if $f: E_1(r) \times E_2(r) \to E$ satisfies $\text{Lip}(f - t) < \varepsilon$ and $\|f(0)\| < \delta$, then $W_r^s(f)$ is the graph of a Lipschitz function $g: E_1(r) \to E_2(r)$ with $\text{Lip}(g) \leq 1$.*

Moreover, if f is C^k then so is g.

Proof. The basic observation is this: for $x \in W_r^s(f)$ the sequence $\gamma = \{\gamma(n) | n \geq 1, \gamma(n) = f^n(x)\}$ satisfies the following conditions: $\|\gamma(n)\| \leq r$, $f(\gamma(n)) - \gamma(n + 1) = 0$, $\forall n \geq 1$ and $f(x) - \gamma(1) = 0$. Conversely, if a point x in $E(r)$ and a sequence $(\gamma(n))_{n \geq 1}$ satisfy these conditions, then x belongs to $W_r^s(f)$ and $\gamma(n) = f^n(x)$. To develop this idea we must introduce some notation.

Let $\mathbf{B} = \{\gamma | \gamma = (\gamma(n))_{n \geq 1}, \gamma(n) \in E, \sup_{n \geq 1} \|\gamma(n)\| < \infty\}$. With the norm $\|\gamma\| = \sup_{n \geq 1} \|\gamma(n)\|$, $\mathbf{B}$ is a Banach space. Let $\mathbf{B}(r)$ be the closed ball of radius r in $\mathbf{B}$. Endow $E \times \mathbf{B}$ and $E_1 \times \mathbf{B}$ with box norms induced by E_1, E, and $\mathbf{B}$.

Now we define a map

$$\mathbf{F}: E_1(r) \times E_2(r) \times \mathbf{B}(r) \to E_1 \times \mathbf{B},$$

$$\mathbf{F}: (x_1, x_2, \gamma) \mapsto (x_1, \mathbf{F}_{x_1}(\gamma)),$$

where $\mathbf{F}_{x_1}: E_2(r) \times \mathbf{B}(r) \to \mathbf{B}$ is defined as

$$\mathbf{F}_{x_1}(\gamma)(1) = f(x_1, x_2) - \gamma(1),$$

$$\mathbf{F}_{x_1}(\gamma)(n) = f(\gamma(n-1)) - \gamma(n) \qquad \text{for} \quad n \geq 2.$$

If we show that $\mathbf{F}$ is invertible and that its image contains $E_1(r) \times 0$, we can define g by $g = \Pi_2 \mathbf{F}^{-1}|_{E_1(r) \times 0}$, where Π_2 is the projection of $E_1 \times E_2 \times \mathbf{B}$ onto E_2. We will, of course, use the Lipschitz inverse function theorem to do this.

Define $\mathbf{T}$ analogously to $\mathbf{F}$ with T in place of f. $\mathbf{T}$ is clearly a continuous linear operator with $\|\mathbf{T}\| \leq 1 + \|\mathrm{T}\|$ and $\mathrm{Lip}(\mathbf{F} - \mathbf{T}) \leq \mathrm{Lip}(f - T)$. Consequently, if $\mathbf{T}$ is invertible and $\mathrm{Lip}(f - T) \leq \|\mathbf{T}^{-1}\|^{-1}$ then $\mathbf{F}$ is also invertible.

T *is invertible.*

$\mathbf{T}$ is given explicitly as $\mathbf{T}(x_1, x_2, \gamma) = (x, v)$ where

$$v(1) = \mathbf{T}_{x_1}(\gamma)(1) = T(x_1, x_2) - \gamma(1),$$

$$v(n) = \mathbf{T}_{x_1}(\gamma(n)) = T(\gamma(n-1)) - \gamma(n), \qquad n \geq 2.$$

We need to express x_2 and γ in terms of x_1 and v. Let us write $\gamma(n) = (\gamma_1(n), \gamma_2(n)) \in E_1 \times E_2$, $v(n) = (v_1(n), v_2(n)) \in E_1 \times E_2$; the definition of v then becomes

$$v_1(1) = T_1(x_1) - \gamma_1(1),$$

$$v_2(1) = T_2(x_2) - \gamma_2(1),$$

$$v_1(n) = T_1(\gamma_1(n-1)) - \gamma_1(n), \qquad n \geq 2,$$

$$v_2(n) = T_2(\gamma_2(n-1)) - \gamma_2(n), \qquad n \geq 2.$$

The first and third equations may be rewritten as

$$\gamma_1(1) = -v_1(1) + T_1(x_1),$$

$$\gamma_1(n) = T_1(\gamma_1(n-1)) - v_1(n), \qquad n \geq 2,$$

which gives

$$\gamma_1(n) = -\left[\sum_{j=1}^{n} T_1^{n-j}(v_1(j))\right] + T_1^n(x_1).$$

The fourth equation gives

$$\begin{aligned}\gamma_2(n-1) &= T_2^{-1}[v_2(n) + \gamma_2(n)] \\ &= T_2^{-1}[v_2(n) + T_2^{-1}(v_2(n+1) + \gamma_2(n+1))] = \cdots\end{aligned}$$

so, in the limit

$$\gamma_2(n) = \sum_{j=1}^{\infty} T_2^{-j}(v_2(n+j)).$$

Finally, from the second equation

$$x_2 = \sum_{j=1}^{\infty} T_2^{-j}(v(j)).$$

These series converge since $\|T_2^{-1}\| < 1$ and $\sup_{n\geq 1} \|\gamma(n)\| < \infty$. Thus, we have found the inverse of $\mathbf{T}$ as

$$\mathbf{T}^{-1}: E_1 \times \mathbf{B} \to E_1 \times E_2 \times \mathbf{B},$$

$$\mathbf{T}^{-1}: (x_1, v) \mapsto (x_1, x_2, \gamma),$$

where

$$x_2 = \sum_{j=1}^{\infty} T_2^{-j}(v_2(j)),$$

$$\gamma_1(n) = T_1^n(x_1) - \sum_{j=1}^{n} T_1^{n-j}(v_1(j)),$$

$$\gamma_2(n) = \sum_{j=1}^{\infty} T_2^{-j}(v_2(n+j)),$$

and one can easily check that $\|\mathbf{T}^{-1}\| \leq (1-\lambda)^{-1}$.

If $\mathrm{Lip}(f - T) < \varepsilon < 1 - \lambda$, *then* $\mathbf{F}$ *is invertible.*

Since $\mathrm{Lip}(\mathbf{F} - \mathbf{T}) \leq \mathrm{Lip}(f - T) < \varepsilon < 1 - \lambda \leq \|\mathbf{T}^{-1}\|^{-1}$, the Lipschitz inverse function theorem applies.

The image of $\mathbf{F}$ *contains* $E_1(r) \times 0$ *when* ε *and* $f(0)$ *are small.*

Note that the image of $\mathbf{F}$ contains $(x_1, 0)$ if and only if the image of $\mathbf{F}_{x_1}$ contains 0. Furthermore, $\mathbf{F}_{x_1}$ is a Lipschitz perturbation of $\mathbf{T}_{x_1}$ and $\mathrm{Lip}(\mathbf{F}_{x_1} - \mathbf{T}_{x_1}) \leq \mathrm{Lip}(f - T)$. Since $\mathbf{T}_{x_1}$ differs from $\mathbf{T}_0$ only by a translation we have, in fact, $\mathrm{Lip}(\mathbf{F}_{x_1} - \mathbf{T}_0) \leq \mathrm{Lip}(f - T)$.

Now since the linear operator $\mathbf{T}_0: E_2 \times \mathbf{B} \to \mathbf{B}$ is given by

$$\mathbf{T}_0(x_2, \gamma)(1) = T(0, x_2) - \gamma(1),$$

$$\mathbf{T}_0(x_2, \gamma)(n) = T(\gamma(n-1)) - \gamma(n), \qquad n \geq 2,$$

we see the $\mathbf{T}_0$ is invertible with inverse

$$\mathbf{T}_0^{-1}: \mathbf{B} \to E_2 \times \mathbf{B},$$

$$\mathbf{T}_0^{-1}: v \to (x, v),$$

where

$$x_2 = \sum_{j=1}^{\infty} T_2^{-j}(v_2(j)),$$

$$\gamma_1(n) = \sum_{j=1}^{n} T_1^{n-j}(v_1(j)),$$

$$\gamma_2(n) = \sum_{j=1}^{\infty} T_2^{-j}(v_2(n+j)),$$

and $\|\mathbf{T}_0^{-1}\| \leq (1-\lambda)^{-1}$.

If $\mathrm{Lip}(f - T) < \varepsilon < 1 - \lambda$, we see that $\mathrm{Lip}(\mathbf{T}_0^{-1}\mathbf{F}_{x_1} - \mathrm{id}) \leq \varepsilon/(1 - \lambda)$ and using the lemma in Appendix I

$$\mathbf{T}_0^{-1}\mathbf{F}_{x_1}[E_2(r) \times \mathbf{B}(r)] \supset \mathbf{T}_0^{-1}\mathbf{F}_{x_1}(0) + E_2(s) \times \mathbf{B}(s),$$

where $s = r(1 - \varepsilon/(1 - \lambda))$.

Now let us calculate $\|\mathbf{T}_0^{-1}\mathbf{F}_{x_1}(0)\|$. First of all $\mathbf{F}_{x_1}(0) = v$, where $v(1) = f(x_1, 0)$ and $v(n) = f(0, 0)$. Consequently, $\mathbf{T}_0^{-1}\mathbf{F}_{x_1}(0) = (x_2, \gamma)$, where

$$x_2 = T_2^{-1}[f_2(x_1, 0)] + \sum_{j=2}^{\infty} T_2^{-j}[f_2(0, 0)],$$

$$\gamma_1(n) = -T_1^{n-1}[f_1(x, 0)] - \sum_{j=2}^{n} T_1^{n-j}[f_1(0, 0)],$$

$$\gamma_2(n) = \sum_{j=1}^{\infty} T_2^{-j}[f_2(0, 0)].$$

Now we can estimate that

$$\begin{aligned} \|f_1(x_1, 0)\| &\leq \|p_1(f - T)(x_1, 0)\| + \|T_1 x_1\| \\ &\leq \|(f - T)(x_1, 0)\| + \lambda\|x_1\|, \\ \|f_2(x_1, 0)\| &\leq \|p_2(f - T)(x_1, 0)\| \leq \|(f - T)(x_1, 0)\|, \end{aligned}$$

and further

$$\begin{aligned} \|(f - T)(x_1, 0)\| &\leq \|(f - T)(0)\| + \|(f - T)(x_1, 0) - (f - T)(0)\| \\ &\leq \|f(0)\| + \varepsilon\|x_1\|. \end{aligned}$$

From this we estimate $\|\mathbf{T}_0^{-1}\mathbf{F}_{x_1}(0)\|$ as follows:

$$\|x_2\| \leq \varepsilon\lambda\|x_1\| + \frac{\lambda}{1 - \lambda}\|f(0)\|,$$

$$\begin{aligned} \|\gamma_1(n)\| &\leq \lambda^n\|x_1\| + \varepsilon\lambda^{n-1}\|x_1\| + \lambda^{n-1}\|f(0)\| + \sum_{j=2}^{n} \lambda^{n-j}\|f(0)\| \\ &\leq \|x_1\|(\lambda + \varepsilon) + \frac{1}{1 - \lambda}\|f(0)\|, \end{aligned}$$

$$\|\gamma_2(n)\| \leq \frac{\lambda}{1 - \lambda}\|f(0)\|,$$

so

$$\|\mathbf{T}_0^{-1}\mathbf{F}_{x_1}(0)\| \leq (\lambda + \varepsilon)r + \frac{1}{1 - \lambda}\|f(0)\|.$$

Since the image of $\mathbf{T}_0^{-1}\mathbf{F}_{x_1}$ contains the ball about $\mathbf{T}_0^{-1}\mathbf{F}_{x_1}(0)$ of radius $s = r(1 - \varepsilon/(1 - \lambda))$, it contains 0 whenever $(\lambda + \varepsilon)r + 1/(1 - \lambda)\|f(0)\| < r(1 - \varepsilon/(1 - \lambda))$. We can rewrite this inequality as

$$\|f(0)\| < (1 - \lambda)r\left(1 - \lambda - \varepsilon - \frac{\varepsilon}{1 - \lambda}\right) = r[(1 - \lambda)^2 - \varepsilon(2 - \lambda)].$$

Suppose then that $\varepsilon < (1-\lambda)^2/(2-\lambda) < 1-\lambda$ and take $\delta > 0$ such that $\delta < r[(1-\lambda)^2 - \varepsilon(2-\lambda)]$; if $\text{Lip}(f-T) < \varepsilon$ and $\|f(0)\| < \delta$, we have shown that $\mathbf{T}_0^{-1}\mathbf{F}_{x_1}$ contains 0. Since $\mathbf{T}_0$ is linear, the image of $\mathbf{F}_{x_1}$ contains 0 as well, so the image of $\mathbf{F}$ contains $E_1(r) \times 0$, as desired.

Looking back over what we have done, we see we have at this point shown that:

If $\text{Lip}(f-T) < \varepsilon < (1-\lambda)^2/(2-\lambda)$ *and* $\|f(0)\| < \delta < r[(1-\lambda)^2 - \varepsilon(2-\lambda)]$, *then* $W_r^s(f)$ *is the graph of a Lipschitz function* $g = \Pi_2 \mathbf{F}^{-1}|_{E_1(r)\times 0}: E_1(r) \to E_2(r)$ *with* $\text{Lip}(g) \leq 1$.

To finish the proof of the stable manifold theorem we only need:

For f above, g is C^k whenever f is C^k.

To show that g is C^k we clearly need only show that $\mathbf{F}^{-1}$ is C^k. Since $\text{Lip}(\mathbf{F}-\mathbf{T}) < \|\mathbf{T}^{-1}\|^{-1}$, if $\mathbf{F}$ is C^1 we have, for all (x, γ) in $E(r) \times \mathbf{B}(r)$,

$$\|D\mathbf{F}_{(x,\gamma)} - \mathbf{T}\| < \|\mathbf{T}^{-1}\|^{-1}.$$

This means that $D\mathbf{F}_{(x,\gamma)}$ is a linear isomorphism for all (x, γ) in $E(r) \times \mathbf{B}(r)$, so the C^k inverse function theorem shows that $\mathbf{F}^{-1}$ is C^k whenever $\mathbf{F}$ is.

We are reduced then to showing that $\mathbf{F}$ is C^k when f is C^k. Unfortunately, $\mathbf{B}$ is too large for this to be possible, but this is a difficulty we can easily surmount.

Notice that the obvious candidate for the derivative of $\mathbf{F}$ at a point $(x, \gamma) = (x_1, x_2, \gamma)$ of $E_1(r) \times E_2(r) \times \mathbf{B}(r)$ is the linear map

$$\mathbf{L}: E_1 \times E_2 \times \mathbf{B} \to E_1 \times \mathbf{B},$$
$$\mathbf{L}: (y_1, y_2, v) \to (y_1, \zeta),$$

where

$$\zeta(1) = Df_x(y) - v(1) \qquad \text{with} \quad y = (y_1, y_2),$$
$$\zeta(n) = Df_{\gamma(n-1)}[v(n-1)] - v(n), \qquad n \geq 2.$$

We have $\mathbf{F}(x+y, \gamma+v) - \mathbf{F}(x, \gamma) - \mathbf{L}(y, v) = (0, \varphi) \in E_1 \times \mathbf{B}_1$, where

$$\varphi(1) = f(x+y) - f(y) - Df_x(y) = \int_0^1 (Df_{x+ty} - Df_x)(y)\, dt,$$

$$\varphi(n) = f[\gamma(n-1) + v(n-1)] - f(\gamma(n-1)) - Df_{\gamma(n-1)}(v(n-1))$$
$$= \int_0^1 (Df_{\gamma(n-1)+tv(n-1)} - Df_{\gamma(n-1)})(v(n-1))\, dt, \qquad n \geq 2.$$

This shows the quotient

$$\frac{\|\mathbf{F}(x+y, \gamma+v) - \mathbf{F}(x, \gamma) - \mathbf{L}(y, v)\|}{\|(y, v)\|}$$

is bounded above by

$$\max\left(\int_0^1 \|Df_{x+ty} - Df_x\|\,dt, \sup_{n\geq 1}\int_0^1 \|Df_{\gamma(n)+tv(n)} - Df_{\gamma(n)}\|\,dt\right).$$

The sup of all those integrals does not necessarily tend to zero with $\|y, v\|$, since **B** is not locally compact and Df is not necessarily uniformly continuous. If we knew that $\{\gamma(n)|n \geq 1\}$ were contained in a compact subset of $E(r)$, in particular, if $\gamma(n)$ converged then the sup would go to zero wih $\|y, v\|$. This leads us to consider the subspace **C** of convergent sequences:

$$\mathbf{C} = \{\gamma \in \mathbf{B}|\lim_{n\to\infty} \gamma(n) \text{ exists}\}.$$

Since E is complete **C** is also the subspace of all Cauchy sequences, which one can easily convince oneself is closed in **B**. **C** itself then is also a Banach space.

Notice that $\mathbf{F}(E_1(r) \times E_2(r) \times \mathbf{C}(r)) \subset E_1(r) \times \mathbf{C}$ so that it makes sense to consider the restriction $\tilde{\mathbf{F}}$ of **F** to $E_1(r) \times E_2(r) \times \mathbf{C}(r)$ and to define $\tilde{\mathbf{T}}$ similarly as the mapping induced by **T**. We have essentially shown that $\tilde{\mathbf{F}}$ is C^1 above and we leave it to the reader to show in general that $\tilde{\mathbf{F}}$ is C^k when F is C^k.

We can now repeat our previous construction of g with $\tilde{\mathbf{F}}$, $\tilde{\mathbf{T}}$, and **C** in place of **F**, **T**, and **B**, provided that we know $\tilde{\mathbf{T}}$ is invertible, i.e., that $\mathbf{T}^{-1}(E_1 \times \mathbf{C}) \subset E_1 \times E_2 \times \mathbf{C}$. This we proceed to check.

Let (x, v) be in $E_1 \times \mathbf{C}$ and recall that $\mathbf{T}^{-1}(x, v) = (x_1, x_2, \gamma)$ is given by

$$x_2 = \sum_{j=1}^{\infty} T_2^{-j}[v_2(j)],$$

$$\gamma_1(n) = T_1^n(x_1) - \sum_{j=1}^{n} T^{n-j}[v_1(j)],$$

$$\gamma_2(n) = \sum_{j=1}^{\infty} T_2^{-j}[v_2(n+j)].$$

To show that γ belongs to **C** we will show that γ_1 and γ_2 are Cauchy. First, γ_1:

$$\|\gamma_1(n) - \gamma_1(m)\| \leq \left\|\sum_{j=1}^{n} T^{n-j}[v_1(j)] - \sum_{j=1}^{m} T^{m-j}[v_1(j)]\right\| + \|T_1^n(x_1) - T_1^m(x_1)\|.$$

Since $\|T_1\| < \lambda < 1$, the term $\|T_1^n(x_1) - T_1^m(x_1)\|$ goes to zero as n and m go to ∞.

As for the other term:

$$\begin{aligned}
&\left\|\sum_{j=1}^{n} T_1^{n-j}[v_1(j)] - \sum_{j=1}^{m} T_1^{m-j}[v_1(j)]\right\| \\
&= \left\|\sum_{k=0}^{n-1} T^k[v_1(n-k)] - \sum_{k=0}^{m-1} T^k[v_1(m-k)]\right\| \\
&\leq \left(\sum_{k=0}^{\infty} \lambda^i\right) \sup_{\substack{h\geq n-N\\ l\geq m-N}} \|v_1(h) - v_1(l)\| + 2\left(\sum_{k=N+1}^{\infty} \lambda^i\right)\|v_1\|.
\end{aligned}$$

For any $\varepsilon > 0$ we can find an N_0 so large that

$$\sup_{\substack{h \geq N_0 \\ l \geq N_0}} \|v_1(h) - v_1(l)\| \leq \frac{1-\lambda}{2}\varepsilon \quad \text{and} \quad \sum_{k=N_0+1}^{\infty} \lambda^i < \frac{\varepsilon}{4\|v_1\|}$$

which tells us that, for n and $m \geq 2N_0$,

$$\left\| \sum_{j=1}^{n} T_1^{n-j}[v_1(j)] - \sum_{j=1}^{m} T_1^{m-j}[v_1(j)] \right\| \leq \left(\sum_{k=0}^{\infty} \lambda^i \right) \frac{1-\lambda}{2}\varepsilon + \frac{2\varepsilon\|v_1\|}{4\|v_1\|} = \varepsilon.$$

We have now shown that γ_1 is Cauchy.

The case of γ_2 is easier, we can estimate

$$\|\gamma_2(n) - \gamma_2(m)\| \leq \left(\sum_{j=1}^{\infty} \lambda^j \right) \sup_{j \geq 1} \|v_2(n+j) - v_2(m+j)\|,$$

so, since the sequence v_2 is Cauchy, so is γ_2.

Thus g can be defined as $g = \tilde{\Pi}_2 \tilde{\mathbf{F}}^{-1}|_{E_1(r) \times 0}$, where $\tilde{\Pi}_2$ is the projection of $E_1 \times E_2 \times \mathbf{C}$ to E_2. Since we have shown above that such a g is as smooth as $\tilde{\mathbf{F}}$ which is as smooth as f, this finishes the proof of the stable manifold theorem. □

Proposition II.5. (i) *If, moreover, an f as above is C^1 and satisfies $f(0) = 0$ and $Df_0 = T$, then $g(0) = 0$ and $Dg_0 = 0$. Consequently, $W_r^s(f)$ is tangent to E_1 at 0.*

(ii) *Let $N_{\varepsilon,\delta}^k = \{f: E_1(r) \times E_2(r) \to E \mid \mathrm{Lip}(f - T) < \varepsilon\}$, $\|f(0)\| < \delta$, f is C^k and $D^k f$ is bounded and uniformly continuous with the C^k topology.*

The map

$$N_{\varepsilon,\delta}^k \to C^k(E_1(r), E_2(r)),$$
$$f \mapsto g$$

is continuous.

PROOF. (i) Surely $g(0) = 0$ since $0 \in W_r^s(f)$. Moreover, $Dg_0 = \tilde{\Pi}_2(D\tilde{\mathbf{F}})_0^{-1}|_{E_1 \times 0}$, but $(D\tilde{\mathbf{F}})_0^{-1} = D\tilde{\mathbf{F}}_0^{-1} = \tilde{\mathbf{T}}^{-1}$, so for v_1 in E_1, we have

$$Dg_0(v_1) = \tilde{\Pi}_2 \tilde{\mathbf{T}}^{-1}(v_1, 0) = \sum_{j=1}^{\infty} T_2^{-j}(0) = 0.$$

(ii) This is a direct consequence of the following facts, whose proof we leave to the reader:

(1) If $f \in N_{\varepsilon,\delta}^k$ then $D^k\tilde{\mathbf{F}}$ is uniformly continuous and bounded and the map $f \to \tilde{\mathbf{F}}$ is continuous in the C^k topology.
(2) The map $\tilde{\mathbf{F}} \to \tilde{\mathbf{F}}^{-1}$ is continuous in the C^k topology on the set of $\tilde{\mathbf{F}}$ whose kth derivative is uniformly continuous and bounded. □

Here is a sketch of another version of Irwin's proof, based on the following two lemmas.

Lemma II.6. *Let $\theta\colon X \times Y \to Y$ be a continuous map. Suppose that Y is a complete metric space and that θ uniformly contracts Y, that is,*

$$\exists k < 1 \forall x \in X, \qquad \forall y, y' \in Y, \qquad d(\theta(x, y)), \theta(x, y') \leq kd(y, y').$$

Let θ_x be the map $y \mapsto \theta(x, y)$ of Y, and φ the map from X to Y which associates to x the fixed point of θ_x. Then φ is continuous, and Lipschitz when X is a metric space and θ is Lipschitz.

Proof. For x and x' in X, we have

$$\begin{aligned} d[\varphi(x), \varphi(x')] &= d[\theta(x, \varphi(x)), \theta(x', \varphi(x'))] \\ &\leq d[\theta(x, \varphi(x)), \theta(x', \varphi(x))] + d[\theta(x', \varphi(x)), \theta(x', \varphi(x'))] \\ &\leq d[\theta(x, \varphi(x)), \theta(x', \varphi(x))] + kd[\varphi(x), \varphi(x')]. \end{aligned}$$

Consequently, we have

$$d[\varphi(x), \varphi(x')] \leq \frac{1}{1-k} d[\theta(x, \varphi(x)), \theta(x', \varphi(x))]$$

and the lemma follows. □

Lemma II.7. *If, moreover, X and Y are balls in some Banach spaces E and F and θ is C^k, then φ is C^k and we have*

$$D\varphi_x = [\mathrm{id} - D_2\theta_{(x,\varphi(x))}]^{-1} D_1\theta_{(x,\varphi(x))}.$$

Proof. First we note that $\|D_2\theta_{(x,y)}\| \leq k < 1$, since $\|\theta(x, y) - \theta(x, y')\| \leq k\|y - y'\|$. Consequently, $\mathrm{id} - D_2\theta_{(x,y)}$ is invertible and $[\mathrm{id} - D_2\theta_{(x,y)}]^{-1} = \sum_{i=0}^{\infty} [D_2\theta_{(x,y)}]^i$.

To prove the lemma for $k = 1$ we need to show that

$$\|\varphi(x + v) - \varphi(x) - [\mathrm{id} - D_2\theta_{(x,\varphi(x))}]^{-1} D_1\theta_{(x,\varphi(x))}(v)\| \text{ is } o(\|v\|).$$

(Recall that $f(x)$ is said to be $o(g(x))$ if $\lim_{x\to 0} f(x)/g(x) = 0$). Well,

$$\begin{aligned} &\|\varphi(x + v) - \varphi(x) - [\mathrm{id} - D_2\theta_{(x,\varphi(x))}]^{-1} D_1\theta_{(x,\varphi(x))}(v)\| \\ &\leq \|[\mathrm{id} - D_2\theta_{(x,\varphi(x))}]^{-1}\| \, \|\varphi(x + v) - \varphi(x) \\ &\quad - D_2\theta_{(x,\varphi(x))}[\varphi(x + v) - \varphi(x)] - D_1\theta_{(x,\varphi(x))}(v)\| \\ &\leq \|[\mathrm{id} - D_2\theta_{(x,\varphi(x))}]^{-1}\| \, \|\theta(x + v, \varphi(x + v)) - \theta(x, \varphi(x)) \\ &\quad - D\theta_{(x,\varphi(x))}[(x + v, \varphi(x + v)) - (x, \varphi(x))]\| \end{aligned}$$

and this last expression is $o(\|(x + v, \varphi(x + v)) - (x, \varphi(x))\|)$, hence $o(\|v\|)$ as well, since φ is Lipschitz.

Thus we have shown that φ is C^1 when θ is C^1, and as usual the rest of the proof (left as an exercise to the reader) follows by induction on k, once we notice that $D\varphi$ can be defined as the composition

$$X \xrightarrow{(\mathrm{id},\varphi)} X \times Y \xrightarrow{(D_1\theta, D_2\theta)} L(E, F) \times U$$
$$\xrightarrow{(\mathrm{id}, p)} L(E, F) \times L(F, F) \xrightarrow{\text{composition}} L(E, F),$$

where E (resp. F) is the Banach space containing X (resp. Y) and $U = \{P \in L(F, F) \mid \|P\| < 1\}$ and $p: P \to [\mathrm{id} - P]^{-1}$. □

Variant of the Proof of II.4. Consider the map

$$\mathbf{G}: E_1(r) \times \mathbf{B}(r) \to \mathbf{B},$$
$$\mathbf{G}: (x_1, \gamma) \mapsto v,$$

where

$$v_1(1) = f_1(x, \gamma_2(1)),$$
$$v_1(n) = f_1(\gamma_1(n-1), \gamma_2(n)), \qquad n \geq 2,$$
$$v_2(1) = T_2^{-1}[T_2\gamma_2(1) + \gamma_2(2) - f_2(x_1, \gamma_2(1))],$$
$$v_2(n) = T_2^{-1}[T_2(\gamma_2(n)) + \gamma_2(n+1) - f_2(\gamma_1(n-1), \gamma_2(n))].$$

Suppose that $\mathrm{Lip}(f - T) < \varepsilon$; an easy calculation shows that

$$\mathrm{Lip}\,\mathbf{G} \leq \max[\lambda + \varepsilon, \lambda(\varepsilon + 1)] = \lambda + \varepsilon.$$

Therefore, if $\varepsilon < 1 - \lambda$, the map $\mathbf{G}$ uniformly contracts the second factor. Supposing that $\|f(0)\| < \delta$, more calculation shows that $\|\mathbf{G}(0)\| < \max(\delta, \lambda\delta) = \delta$. Consequently, $\mathbf{B}(r) \subset \mathbf{G}(E_1(r) \times \mathbf{B}(r))$ whenever $\delta + (\lambda + \varepsilon)r \leq r$, that is, $\delta \leq r(1 - \lambda - \varepsilon)$.

Thus, if $\mathrm{Lip}(f - T) < \varepsilon < 1 - \lambda$ and $\|f(0)\| < \delta < r(1 - \lambda - \varepsilon)$ we can apply Lemma II.6 to $\mathbf{G}$ and find a Lipschitz map $\Phi: E_1(r) \to \mathbf{B}(r)$, such that $\Phi(x_1)$ is the fixed point of G_{x_1}.

Notice that if γ is the fixed point of G_{x_1} and we let $x_2 = \gamma_2(1)$, then $f_1(x_1, x_2) = \gamma_1(1)$, and also

$$\gamma_2(1) = T_2^{-1}[T_2(\gamma_2(1)) + \gamma_2(2) - f_2(x_1, \gamma_2(1))],$$

that is, $\gamma_2(2) = f_2(x_1, \gamma_2(1)) = f_2(x_1, x_2)$.

By induction we obtain, with a slight abuse of notation, $f^n(x_1, x_2) = (f_1^n(x_1, x_2), f_2^n(x_1, x_2)) \in E_1(r) \times E_2(r)$, $\gamma_1(n) = f_1^n(x_1, x_2)$ and $\gamma_2(n+1) = f_2^n(x_1, x_2)$. Therefore, if we define $g: E_1(r) \to E_2(r)$ by $g = \Pi\Phi$, where $\Pi: \mathbf{B} \to E_2$, $\gamma \mapsto \gamma_2(1)$, then $W_r^s(f)$ is precisely the graph of g. This finishes the theorem in the Lipschitz case.

For the C^k case, Lemma II.7 shows that if $\mathbf{G}$ is C^k then g is C^k. As before, though, $\mathbf{G}$ is not necessarily C^k when f is C^k, so we must again shift our attention from $\mathbf{B}$ to $\mathbf{C}$. We leave the details to the reader. □

Appendix III

Center and Strong Stable Manifolds

Here we indicate how proofs of generalizations of the stable manifold theorem proceed. Ultimately we will be concerned with center, center stable and unstable, and strong stable and unstable manifolds.

Definition III.1. Let $T: E \to E$ be a continuous linear map of the Banach space E. **T** is ρ-pseudohyperbolic if there is a **T** invariant direct sum decomposition $E = E_1 \oplus E_2$ and constants $0 < \lambda_1 < \rho < \mu_1$ and $C_1, C_2 > 0$ such that:

(1) the restriction T_1 of T to E_1 is an isomorphism and $\forall n \geq 0$ and $\forall v \in E_1$

$$\|T_1^n(v)\| \geq C_1 \mu_1^n \|v\|;$$

(2) $\forall n \geq 0$ and $\forall v \in E_2$ and T_2 the restriction of T to E_2

$$\|T_2^n(v)\| \leq C_2 \lambda_1^n \|v\|.$$

A pseudohyperbolic linear map is hyperbolic when $\lambda_1 < 1 < \mu_1$. By renorming E we may assume that the norm is adapted to **T**, that is, $\|(x, y)\| = \max(\|x\|, \|y\|)$ and there are $0 < \lambda < \rho < \mu$ such that

(1) $$\|T_1(v)\| > \mu \|v\| \quad \text{for all} \quad v \neq 0 \in E_1,$$

(2) $$\|T_2(v)\| < \lambda \|v\| \quad \text{for all} \quad v \neq 0 \in E_2.$$

Theorem III.2. *Let $T: E \to E$ be a ρ-pseudohyperbolic continuous linear map of the Banach space E, with splitting $E = E_1 \oplus E_2$, adapted metric $\| \ \|$ and constants $0 < \lambda < \rho < \mu$ such that*

$$\|T_1(v)\| > \mu \|v\| \quad \text{for all} \quad v \neq 0 \text{ in } E_1,$$

$$\|T_2(v)\| < \lambda \|v\| \quad \text{for all} \quad v \neq 0 \text{ in } E_2.$$

There is a real number $\varepsilon > 0$ such that if $f: E \to E$ is a Lipschitz map with $f(0) = 0$ and $\mathrm{Lip}(f - T) < \varepsilon$, then

(1) *The set $W_1 = \bigcap_{n \geq 0} f^n S_1$, where $S_1 = \{(x, y) \in E_1 \times E_2 : \|x\| \geq \|y\|\}$ is the graph of a Lipschitz function g of Lipschitz constant less than or equal to one, $g: E_1 \to E_2$ and f maps the graph of g to itself.*
(2) *$z \in W_1$ if and only if there are inverse images $f^{-n}z$ such that $\|f^{-n}(z)\|/\rho^n \to 0$ as $n \to \infty$ or even $\|f^{-n}(z)\|/\rho^n$ stays bounded as $n \to \infty$.*
(3) *If f is C^r and $\mu^{-j}\lambda < 1$ for $1 \leq j \leq r$ then g is C^r. If f is differentiable at 0 and if $Df(0) = T$ then the graph of g is tangent to E_1 at 0.*

For $\mu < 1$, the graph of g is called the center unstable manifold and denoted by W^{cu} or perhaps $W_f^{cu}(0)$ if some confusion is possible. W^{cu} might actually even contain some part of the stable manifold, that is, points which are asymptotic to 0 under iteration of f. If $\lambda > 1$ then the graph of g is the strong unstable manifold, W^{uu}, and f may contain other unstable points outside of W^{uu} but asymptotic to 0 under iterates of f^{-1}.

If f is invertible then by considering f^{-1} there is an invariant manifold tangent to E_2 at 0, which is the intersection $\bigcap_{n\geq 0} f^{-n}S_2$, where $S_2 = \{(x, y) \in E_1 \times E_2 \mid \|x\| \leq \|y\|\}$, etc. These manifolds are denoted W^{cs} if $\lambda > 1$ and W^{ss} if $\mu < 1$. So if $1 < \lambda < \mu$ we have $W^{cs} \pitchfork W^{uu}$ at 0 which is the only point of intersection; if $0 < \lambda < \mu < 1$ then $W^{ss} \pitchfork W^{cu}$ at 0 and 0 is the only point of intersection; if $0 < \lambda < 1 < \mu$ we are back in the usual hyperbolic case and $W^s \pitchfork W^u$ at 0. Recall that $\pitchfork$ means intersects transversally.

The proof of this theorem, although rather similar to the unstable manifold theorem has a point or two of difference.

Consider functions $g: E_1 \to E_2$ such that $g(0) = 0$ and $\text{Lip}(g) \leq 1$. The graph transform $\Gamma_f(g) = f_2 \circ (\text{id}, g) \circ h^{-1}$, where $h = f_1 \circ (\text{id}, g)$ and h^{-1} is Lipschitz with Lipschitz constant $\leq 1/(\mu - \varepsilon)$ as in Lemma 5.5. Γ_f is not necessarily contracting in the sup norm. We define a new metric

$$\|g_1 - g_2\|_* = \sup_{x \neq 0} \frac{\|g_1(x) - g_2(x)\|}{\|x\|}, \qquad x \in E_1.$$

Lemma III.3. *With the norm* $\| \ \|_*$ *the space* $G = \{g: E_1 \to E_2 \mid g(0) = 0$ *and* $\|g\|_* < \infty\}$ *is a Banach space and*

$$G(1) = \{g \in G: \text{Lip } g \leq 1\}$$

is a closed subset.

Proof. If g_n is a Cauchy sequence, the g_n converge uniformly on bounded sets and hence pointwise to a function g. Moreover, $\|g_n\|_*$ is bounded by some number k, thus for each x, $\|g_n(x)\|/\|x\| \leq k$ and the same is true for $g, \|g(x)\|/\|x\| \leq k$ for all $x \in E_1$. Given $\varepsilon > 0$, there is an N such that for n, $m > N$, $\|g_n - g_m\|_* = \sup_{x\neq 0} \|g_n(x) - g_m(x)\|/\|x\| < \varepsilon/2$. For any particular x choose m so large that $\|g_m(x) - g(x)/\|x\| < \varepsilon/2$ and we have

$$\frac{\|g(x) - g_n(x)\|}{\|x\|} \leq \frac{\|g(x) - g_m(x)\|}{\|x\|} + \frac{\|g_m(x) - g_n(x)\|}{\|x\|} < \varepsilon.$$

Thus, for $n > N$, $\|g - g_n\|_* < \varepsilon$ and g_n converges to g. Finally, as convergence in $\| \ \|_*$ implies uniform convergence on bounded sets, we see that if $\text{Lip}(g_n) \leq 1$ then $\text{Lip}(g) \leq 1$ as well, so $G(1)$ is closed in G.

For any $g \in G(1)$, $\Gamma_f(g)$ is defined and $\text{Lip } \Gamma_f(g) \leq 1$, in fact $\text{Lip } \Gamma_f(g) \leq \text{Lip } f_2 \text{ Lip}(\text{Id}, g)\text{Lip } h^{-1} \leq (\lambda + \varepsilon)/(\mu - \varepsilon) < 1$ for ε small enough. So we have shown:

Lemma III.4. $\Gamma_f: G(1) \to G(1)$.

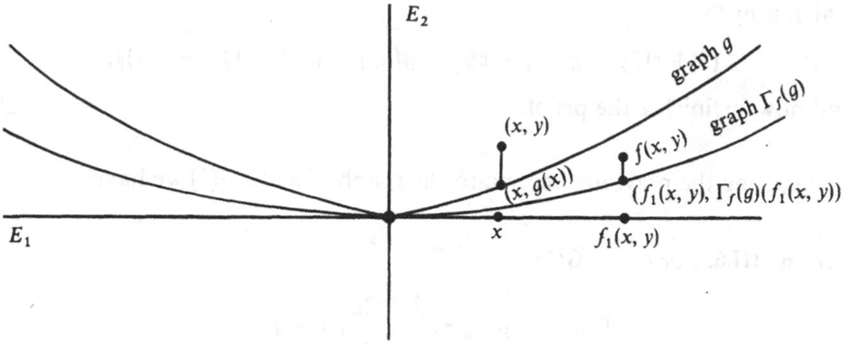

Figure III.1.

Lemma III.5. If $\|x\| \geq \|y\|$ and $g \in G(1)$ then

$$\frac{\|f_2(x, y) - \Gamma_f(g)(f_1(x, y))\|}{\|f_1(x, y)\|} < \frac{\lambda + 2\varepsilon}{\mu - \varepsilon} \frac{\|y - g(x)\|}{\|x\|}.$$

See Figure III.1.

PROOF. (a)

$$\begin{aligned}\|f_2(x, y) - f_2(x, g(x))\| &\leq \|(T - f)_2(x, y) - (T - f)_2(x, g(x))\| \\ &\quad + \|T_2(x, g) - T_2(x, g(x))\| \\ &< \varepsilon\|y - g(x)\| + \lambda\|y - g(x)\| \\ &< (\lambda + \varepsilon)\|y - g(x)\|;\end{aligned}$$

(b)

$$\|f_1(x, y) - f_1(x, g(x))\| \leq \varepsilon\|y - g(x)\|$$

similarly

(c)

$$\|f_1(x, y)\| = \|T_1(x, y) + (f_1 - T_1)(x, y)\| \geq \|T_1(x, y)\| - \|(f_1 - T_1)(x, y)\|.$$

Now, since $\|x\| \geq \|y\|$ and $f(0) = 0$, this last is $\geq \mu\|x\| - \varepsilon\|x\| = (\mu - \varepsilon)\|x\|$. Thus

$$\begin{aligned}\|f_2(x, y) - \Gamma_f(g)(f_1(x, y))\| &\leq \|f_2(x, y) - f_2(x, g(x))\| \\ &\quad + \|f_2(x, g(x)) - \Gamma_f(g)f_1(x, y)\|\end{aligned}$$

which by (a)

$$\begin{aligned}&\leq (\lambda + \varepsilon)\|y - g(x)\| + \|\Gamma_f(g)(f_1(x, g(x))) - \Gamma_f(g)(f_1(x, y))\| \\ &\leq (\lambda + \varepsilon)\|y - g(x)\| + \text{Lip}\,\Gamma_f(g)\|f_1(x, g(x)) - f_1(x, y)\|\end{aligned}$$

which is by (b)

$$\leq (\lambda + \varepsilon)\|y - g(x)\| + \varepsilon\|y - g(x)\| = (\lambda + 2\varepsilon)\|y - g(x)\|,$$

and now (c) finishes the proof. □

Applying the previous Lemma to the graph of a $g' \in G(1)$ we have:

Lemma III.6. *For* $g, g' \in G(1)$

$$\|\Gamma_f g - \Gamma_f g'\|_* \leq \frac{\lambda + 2\varepsilon}{\mu - \varepsilon}\|g - g'\|_*.$$

We are now ready for the proof of the Lipschitz version of Theorem III.2. Let $g \in G(1)$ be the fixed point of Γ_f. Graph $g \subset W_1$ and $f(\text{graph } g) = \text{graph } g$. If $(x, y) \in \text{graph } g$ then $f^n(h^{-n}(x), gh^{-n}(x)) = (x, y)$, where $h = f_1 \circ (\text{id}, g)$ and

$$\|(h^{-n}(x), gh^{-n}(x))\| = \|(h^{-n}(x))\| \leq \left(\frac{1}{\mu - \varepsilon}\right)^n \|x\|$$

since $h^{-1}(0) = 0$ and Lip $h^{-1} < 1/(\mu - \varepsilon)$. Thus

$$\|(h^{-n}(x), gh^{-n}(x))\|/\rho^n < \left(\frac{\rho}{\mu - \varepsilon}\right)^n \|x\|$$

which tends to zero for ε small.

If (x, y) is any point in S_1, $\|y - g(x)\|/\|x\| \leq 2$. If $(x_n, y_n) \in S_1$ and $f^n(x_n, y_n) = (x, y)$ then

$$\frac{\|y_n - g(x_n)\|}{\|x_n\|} > \left(\frac{\mu - \varepsilon}{\lambda + 2\varepsilon}\right)^n \frac{\|y - g(x)\|}{\|x\|}$$

by Lemma III.5. Thus, if (x, y) is not on the graph of g there is an n such that $f^{-n}(x, y) \subset S_2 = \{(x, y) \in E_1 \times E_2 \mid \|y\| \geq \|x\|\}$. This proves Theorem III.2.1 and half of III.2.2. What remains of III.2.2 is to show that if $(x, y) \neq (0, 0) \in S_2$ and $(x_m, y_m) \in f^{-m}(x, y)$ then $\|(x_m, y_m)\|/\rho_m \to \infty$ as $m \to \infty$. Since $\|f_1(x, y)\| \geq (\mu - \varepsilon)\|x\|$ on S_1 and $\|f_2(x, y)\| \leq (\lambda + \varepsilon)\|y\| + \varepsilon\|x\|$, $f\colon S_1 \to S_1$ and $f^{-1}(S_2) \subset S_2$. Moreover, for $(x, y) \in S_2$ and for $(x', y') \in f^{-1}(x, y)$

$$\frac{1}{\lambda + 2\varepsilon}\|(x, y)\| < \|(x', y')\|.$$

Thus,

$$\|(x_m, y_m)\| > \left(\frac{1}{\lambda + 2\varepsilon}\right)^m \|(x, y)\|$$

$$\text{and} \quad \|(x_m, y_m)\|/\rho^m > \left(\frac{\rho}{\lambda + 2\varepsilon}\right)^m \|(x, y)\| \to \infty.$$

This concludes the Lipschitz part of Theorem III.2. Now for the C^r theorem. Define the bundle $E_1 \times L(E_1, E_2) \to E_1$ as in the proof of the unstable manifold theorem, Theorem 5.2 and proceed as there. In fact, if $\operatorname{Lip}(f - T) < \varepsilon$ then $\|Df(x, y) - T\| < \varepsilon$ for all (x, y) and hence Γ_{Df} is a contraction with contraction constant $(\lambda + 2\varepsilon)/(\mu - \varepsilon)$. By Lemma III.5 $\|\Gamma_{Df}(0)\| < \varepsilon$ so, for ε small enough, $\Gamma_{Df}\colon L_1(E, F) \to L_1(E, F)$ and the map F

$$\begin{array}{ccc} E_1 \times L_1(E_1, E_2) & \xrightarrow{F} & E_1 \times L_1(E_1, E_2) \\ \big\downarrow & & \big\downarrow \\ E_1 & \xrightarrow{h} & E_1, \end{array}$$

defined by $F(x, L) = (h(x), \Gamma_{Df}L)$ is a fiber contraction with $\operatorname{Lip}(h^{-1}) \leq 1/(\mu - \varepsilon)$ and fiber contraction constant $\leq (\lambda + 2\varepsilon)/(\mu - \varepsilon)$. For $r \geq 2$, if we know that g is C^{r-1} then F is C^{r-1} and the C^r Section Theorem 5.18 proves that g is C^r by concluding that the tangent bundle to the graph of g is C^{r-1}. The proof is reduced to showing that g is C^1. The map Γ_F still has a unique fixed section σ which is the candidate for the derivative of g and which is 0 at 0 if $Df(0) = \mathbf{T}$. Now proceed exactly as Proposition 5.16 except that (*) is replaced by

$$\begin{aligned} &\operatorname{Lip}_0[(\Gamma_f g)(h(x) + y), gh(x) + \Gamma_{Df}(\pi_2\sigma(x))y] \\ (*) \qquad &\leq \left(\frac{\lambda + 2\varepsilon}{\mu - \varepsilon}\right) \operatorname{Lip}_0[g(x + y), g(x) + \pi_2\sigma(x)y], \end{aligned}$$

since in the estimate of (2) $\operatorname{Lip} k^{-1} < 1/(\mu - \varepsilon)$, $\|w'\| < 1/(\mu - \varepsilon)\|y\|$ and applying Lemma III.5 to $Df_{(x,y)}$ gives the estimate. The term (1) still contributes 0. This proves that g is C^1. If f is simply differentiable at 0 with $Df(0) = \mathbf{T}$ then the above discussion is valid at 0. $\Gamma_{Df}(0)\colon L_1(E_1, E_2) \to L_1(E_1, E_2)$ has the zero function as its unique fixed point and it is tangent to the fixed point of Γ_f at 0. This finishes the proof of the theorem. □

For a linear map $L\colon \mathbb{R}^n \to R^n$ we may always split $\mathbb{R}^n$ into three invariant subspaces $E^s \oplus E^c \oplus E^u$, where E^s, E^c, and E^u are the generalized eigenspaces of eigenvalues of absolute value less than one, equal to one, and greater than one, respectively. Thus there is an adapted norm on $\mathbb{R}^n$ such that L contracts vectors in E^s, expands vectors in E^u, and has no exponential effect on vectors in E^c.

Theorem III.7 (Center and Stable Manifolds). *Let 0 be a fixed point for the C^r local differomorphism $f\colon U \to \mathbb{R}^n$ where U is a neighborhood of zero in $\mathbb{R}^n$ and $\infty > r \geq 1$. Let $E^s \oplus E^c \oplus E^u$ be the invariant splitting of $\mathbb{R}^n$ into the generalized eigenspaces of $Df(0)$ corresponding to eigenvalues of absolute value less than one, equal to one, and greater than one. To each of the five $Df(0)$ invariant subspaces E^s, $E^s \oplus E^c$, E^c, $E^c \oplus E^u$,* and *E^u there is associated a local f invariant*

C^r embedded disc W^s_{loc}, W^{sc}_{loc}, W^c_{loc}, W^{cu}_{loc}, and W^u_{loc} tangent to the linear subspace at 0 and a ball B around zero in an adapted norm such that:

(1) $W^s_{loc} = \{x \in B \mid f^n(x) \in B$ *for all* $n \geq 0$ *and* $d(f^n(x), 0)$ *tends to zero exponentially*$\}$. $f\colon W^s_{loc} \to W^s_{loc}$ *and is a contraction mapping.*
(2) $f(W^{cs}_{loc}) \cap B \subset W^{cs}_{loc}$. *If* $f^n(x) \in B$ *for all* $n \geq 0$, *then* $x \in W^{cs}_{loc}$.
(3) $f(W^c_{loc}) \cap B \subset W^c_{loc}$. *If* $f^n(x) \in B$ *for all* $n \in Z$, *then* $x \in W^c_{loc}$.
(4) $f(W^{cu}_{loc}) \cap B \subset W^{cu}_{loc}$. *If* $f^n(x) \in B$ *for all* $n \leq 0$, *then* $x \in W^{cu}_{loc}$.
(5) $W^u_{loc} = \{x \in B \mid f^n(x) \in B$ *for all* $n \leq 0$ *and* $d(f^n(x), 0)$ *tends to zero exponentially*$\}$. $f^{-1}\colon W^u_{loc} \to W^u_{loc}$ *is a contraction mapping.*

PROOF. Let φ_1 be a standard C^∞ bump function, one on the ball of radius 1 and zero in the complement of the ball of radius 2 around zero in $\mathbb{R}^n$, and $\varphi_s(x) = \varphi_1(x/s)$. Let $f = Df(0) + h$ in U. For s small enough let $h_s = \varphi_s h$ in the ball of radius $2s$ around zero and zero otherwise. Then $f = Df(0) + h_s$ in the ball of radius s. In the ball of radius $2s$, $Dh_s = (D\varphi_s)h + \varphi_s Dh = (1/s)(D\varphi_1)h + \varphi_s Dh$. As $Dh = 0$ at 0, $(1/s)h$ tends to zero and $Dh_s \to 0$ as $s \to 0$. Thus for s sufficiently small we apply Theorem III.2 to $Df(0) + h_s$ to prove (1), (2), (4), and (5). W^c_{loc} is defined as the intersection of W^{sc}_{loc} and W^{cu}_{loc}. Since these discs are tangent to E^{sc} and E^{cu}, respectively, at 0 they intersect transversally in a C^r disc near 0. □

The rates of contraction can be made more explicit via III.2. While the stable and unstable manifolds W^s_{loc} and W^u_{loc} are unique the center stable, center, and center unstable W^{sc}_{loc}, W^c_{loc}, and W^{cu}_{loc} need not be. A simple example is given by the solution curves (Figure III.2.) of the differential equation

$$\dot{x} = \lambda x \quad \text{for} \quad \lambda < 0,$$
$$\dot{y} = y^2,$$

and the time one map near zero.

A form of local uniqueness can be concluded for W^{sc}_{loc}, W^c_{loc}, and W^{cu} under special conditions. If, for example, there is a disc $D \subset W^{sc}_{loc}$ such that $f^n(x) \in B$ for all $x \in D$ and all $n \geq 0$ then D must be contained in any W^{sc}_{loc} by 2.

A version of Theorem III.7 is valid for more general three-way splittings and in Banach space.

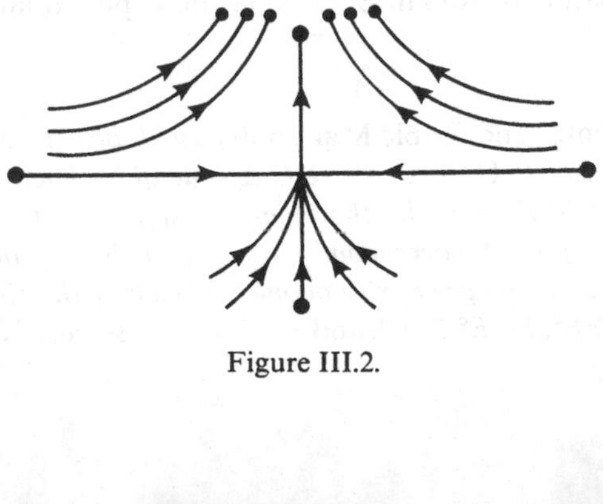

Figure III.2.

Theorem III.8. *Let 0 be a fixed point for the C^r local diffeomorphism $f\colon U \to E$, where U is a neighborhood of zero in the Banach space E and $\infty > r \geq 1$. Suppose that E has a $Df(0)$ invariant direct sum decomposition $E^s \oplus E^c \oplus E^u$ with the max norm and that there are positive real numbers $\lambda_0, \lambda, \lambda_1, \mu_1, \mu, \mu_0$ with $0 < \lambda_0 < \lambda < \lambda_1 < 1 < \mu_1 < \mu < \mu_0$ and*

$$\|Df(0)(v)\| < \lambda_0 \|v\| \quad \text{for} \quad v \neq 0 \quad \text{and} \quad v \in E^s,$$

$$\lambda_1 \|v\| < \|Df(0)(v)\| < \mu_1 \|v\| \quad \text{for} \quad v \neq 0 \quad \text{and} \quad v \in E^c,$$

$$\mu_0 \|v\| < \|Df(0)(v)\| \quad \text{for} \quad v \neq 0 \quad \text{and} \quad v \in E^u.$$

Let $B^s(t)$, $B^c(t)$, and $B^u(t)$ denote the ball of radius $t > 0$ around 0 in E^s, E^c, and E^u, respectively, and $B(t)$ be the product of the three. To each of the five invariant subspaces E^s, $E^s \oplus E^c$, E^c, $E^c \oplus E^u$, and E^u of $Df(0)$, *there is a local f invariant graph defined on a ball of radius $t > 0$.* With W^{ss}_{loc}, W^{sc}_{loc}, W^{c}_{loc}, W^{cu}_{loc}, and W^{uu}_{loc} *tangent to the linear subspace at zero and such that:*

(1) $W^{ss}_{loc} = \{x \in B(t) \mid f^n(x) \in B(t)$ *for all* $n \geq 0$ *and* $d(f^n(x), 0)\lambda^{-n} \to 0$ *as* $n \to \infty\}$. *If $\lambda_0 \lambda_1^{-j} < 1$ for $1 \leq j \leq r$ then W^{ss}_{loc} is the graph of a C^r function $g^{ss}\colon B^s(t) \to B^c(t) \times B^u(t)$ and is tangent to E^s at 0. $f\colon W^{ss}_{loc} \to W^{ss}_{loc}$ is a contraction mapping.*

(2) *W^{sc}_{loc} is the graph of a Lipschitz function $g^{sc}\colon B^s(t) \times B^c(t) \to B^u(t)$ and is tangent to $E^s \oplus E^c$ at 0. $f\colon W^{sc}_{loc} \cap B(t) \subset W^{sc}_{loc}$. If $f^n(x) \in B(t)$ for all $n \geq 0$ then $x \in W^{sc}_{loc}$.*

(3) *W^{c}_{loc} is the graph of a Lipschitz function $g^{c}\colon B^c(t) \to B^s(t) \times B^u(t)$ and is tangent to E^c at 0. $f(W^{c}_{loc}) \cap B(t) \subset W^{c}_{loc}$. If $f^n(x) \in B(t)$ for all $n \geq Z$ then $x \in W^{c}_{loc}$.*

(4) *W^{cu}_{loc} is the graph of a Lipschitz function $g^{cu}\colon B^c(t) \times B^u(t) \to B^s(t)$ and is tangent to $E^c \oplus E^u$ at 0. $f(W^{cu}_{loc}) \cap B(t) \subset W^{cu}_{loc}$. If $f^n(x) \subset B(t)$ for all $n \leq 0$ then $x \in W^{cu}_{loc}$.*

(5) $W^{uu}_{loc} = \{x \in B(t) \mid f^n(x) \in B(t)$ *for all* $n \leq 0$ *and* $d(f^n(x), 0)\mu^n \to 0$ *as* $n \to -\infty\}$. *If $\mu_0^{-1} \mu_1^j < 1$ for $1 \leq j \leq r$ then W^{uu}_{loc} is the graph of a C^r function $g^{uu}\colon B^u(t) \to B^s(t) \times B^c(t)$ and is tangent to E^u at 0. $f^{-1}\colon W^{uu}_{loc} \to W^{uu}_{loc}$ is a contraction mapping.*

Among the following conditions:

(a) *$\mu_0^{-1} \mu_1^j < 1$ for $1 \leq j \leq r$.*
(b) *$\lambda_0 \lambda_1^{-j} < 1$ for $1 \leq j \leq r$.*
(c) *f has C^r extensions f_s such that $f_s = f$ on the ball of radius s, f_s is globally defined on E and converges C^1 to $Df(0)$ as $s \to 0$, for example if B has C^r bump functions, is Hilbert, or is finite dimensional.*
(d) *For some W^{sc}_{loc}, $f^{-1} W^{sc}_{loc} \supset W^{sc}_{loc}$.*
(e) *For some W^{cu}_{loc}, $f(W^{cu}_{loc}) \supset W^{cu}_{loc}$.*
(f) *For some W^{c}_{loc}, $f(W^{c}_{loc}) = W^{c}_{loc}$.*

(i) *If* (a) *and* (c) *or* (a) *and* (d) *hold g^{sc} and hence W^{sc} may be chosen C^r.*
(ii) *If* (b) *and* (c) *or* (b) *and* (e) *hold g^{cu} and hence W^{cu} may be chosen C^r.*
(iii) *If* (a), (b), *and* (c) *or* (a), (b), *and* (f) *hold g^{c} and hence W^{c} may be chosen C^r.*

PROOF. For any Banach Space E there are always Lipschitz bump functions. Let $\varphi: R \to [0, 1]$ be C^∞, identically 1 on $[-1, 1]$ and zero in the complement of $[-2, 2]$. Then $\psi(x) = \varphi(\|x\|)$ defines a bump function on E. Let $\psi_s(x) = \psi((1/s)(x))$ and $f = Df(0) + h$ in a neighborhood of zero. Let

$$h_s = \begin{cases} \psi_s h & \text{on } B(2s), \\ 0 & \text{outside } B(2s), \end{cases}$$

and define $f_s = Df(0) + h_s$. Then $f_s(0) = 0$ and $\text{Lip}(f_s - Df(0)) = \text{Lip}(h_s)$. Now for $x, y \in B(2s)$

$$\begin{aligned} \|h_s(x) - h_s(y)\| &= \|\psi_s(x)h(x) - \psi_s(y)h(y)\| \\ &\leq \psi_s(x)\|h(x) - h(y)\| + \|h(y)\|\,\|\psi_s(x) - \psi_s(y)\| \\ &\leq \psi_s(x)\text{Lip}(h|B(2s)) + \|h(y)\|\,\text{Lip}(\psi_s)\,\|x - y\| \\ &\leq (\text{Lip}(h|B(2s)) + \|h(y)\|\,(1/s)\text{Lip}\,\psi\,\|x - y\|. \end{aligned}$$

If $x \in B(2s)$ and y is not then $\|h_s(x) - h_s(y)\| = \|h_s(x)\| = \|h_s(x) - h_s(y')\|$, where y' is on the line segment between x and y and on the boundary of $B(2s)$, thus $\|x - y'\| \leq \|x - y\|$ and we have shown that

$$\text{Lip}\, h_s \leq \text{Lip}(h|B(2s)) + 1/s \sup_{y \in B(2s)} \|h(y)\|\,\text{Lip}\,\psi.$$

$\text{Lip}(h|B(2s)) \to 0$ as $s \to 0$ since $Dh(0) = 0$ and h is C^1, and $(1/s)\sup_{y \in B(2s)} \|h(y)\| \to 0$ simply because $Dh(0) = 0$.

For s sufficiently small, Theorem III.2 is applied to produce the Lipschitz version of (1), (2), (4), and (5).

$$g^c: B^c(s) \to B^s(s) \times B^u(s) = (g^{sc}|0 \times B^c(s) \times (g^{cu}|B^c(s) \times 0)$$

which proves (3) for the Lipschitz case as well. If the extensions f_s are C^r then Theorem III.2 gives the C^r result too.

If $f: W^{cu}_{loc} \supseteq W^{cu}_{loc}$ or $f^{-1}: W^{sc}_{loc} \supseteq W^{sc}_{loc}$ then the C^r section theorem as in the case of the unstable manifold theorem can be used to prove first the C^1 and then the C^r result. As

$$f(W^{uu}_{loc}) \supseteq W^{uu}_{loc} \quad \text{and} \quad f^{-1}(W^{ss}_{loc}) \supseteq W^{ss}_{loc}.$$

The C^r result follows for these manifolds. If W^{cu}_{loc} and W^{sc}_{loc} are both C^r so is W^c_{loc} and this finishes the proof. □

EXERCISE III.1. Carry out the details of Theorem III.2.

EXERCISE III.2. It is not necessary to assume that f is invertible to conclude the existence of the manifolds W^{ss} or W^{cs} tangent to E_2 in Theorem III.2.

Let G be the Banach space of functions $g: E_2 \to E$ with

$$g(0) = 0 \quad \text{and} \quad \|g\|_* = \sup_{y=0} \frac{\|\sigma(y) - y\|}{\|y\|} < \infty.$$

Let $G(1) = \{g \in G | \text{Lip}\, g \leq 1\}$.

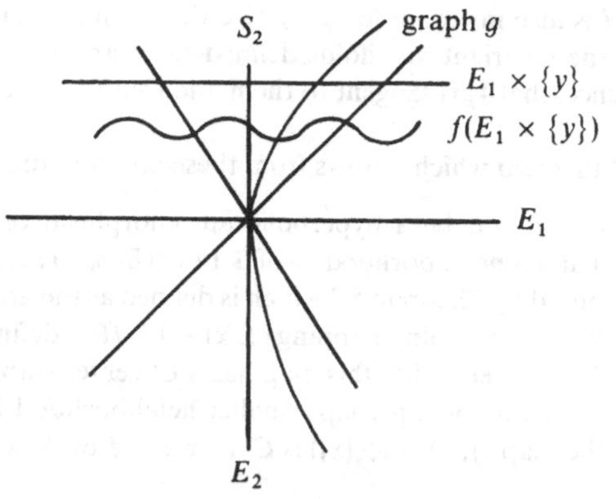

Figure III.3.

(a) Show that a graph transform for f^{-1} is defined on this space for ε small enough, i.e.,

$$\Gamma_{f^{-1}}: G(1) \to G(1), \qquad \text{where} \quad f^{-1}(\text{graph } g) = \text{graph } \Gamma_{f^{-1}} g.$$

To show that $\Gamma_{f^{-1}}$ is defined is like the implict function theorem because for each $y \in E_2$ we are to find a unique point $(x, y) \in E_1 \times \{y\}$ such that $\|x\| \le \|y\|$ and $f(x, y) \in$ graph g, that is, $f(E_1 \times \{y\} \cap S_2)$ should intersect the graph of g in a unique point (Figure III.3). Consider the transformation

$$x \to T_1^{-1}(g(f_2(x, y)) - (f_1 - T_1)(x, y))$$

for $\{x \in E_1 : \|x\| \le \|y\|\}$. Show that this transformation is a contraction of this set into itself and that its fixed point is the unique point sought. Verify that the $\Gamma_{f^{-1}}(g) \in G(1)$.

(b) Show that $\Gamma_{f^{-1}}: G(1) \to G(1)$ is a contraction for ε small enough.

(c) Prove the Lipschitz and C^1 versions of the theorem, having defined $\Gamma_{Df^{-1}}$ and $\Gamma_{F^{-1}}$ on the space of sections of $E_2 \times L_1(E_2, E_1)$.

(d) State and prove the appropriate C^r section theorem to conclude the C^r version of the theorem.

EXERCISE III.3. Generally, the center and stable manifold theorems have flow invariant analogues for vector fields. Given a C^r factor field V defined in a neighborhood of 0 in $\mathbb{R}^n$ such that $V(0) = 0$, let $\varphi_s = \varphi_1(1/s)$ be a standard bump function and

$$V_s = \varphi_s V, \qquad DV_s(0) = DV(0) \qquad \text{for all} \quad s > 0.$$

$\mathbb{R}^n$ has a three-way $DV(0)$ invariant splitting $E^s \oplus E^c \oplus E^u$, where E^s, E^c, and E^u are the generalized eigenspace of the eigenvalues with real part less than 0, equal to 0, and greater than 0, respectively. Let $\psi_s(t)$ be the flow of V_s, and $L(t)$ be the flow of $DV_s(0)$.

(a) Prove that for fixed t

$$\text{Lip}(\psi_s(t) - L(t)) \to 0, \qquad \text{as} \quad s \to 0.$$

(b) Apply Theorem III.2 to $\psi_s(t)$ for small s.

(c) Given an invariant manifold $M \subseteq \mathbb{R}^n$ for $\psi_s(t)$, $\psi_s(t)M = M$, then

$$\begin{aligned}\psi_s(t)\psi_s(t')M &= \psi_s(t')\psi_s(t)M \\ &= \psi_s(t')M\end{aligned}$$

so that $\psi_s(t')M$ is also invariant for $\psi_s(t)$. Use the uniqueness in Theorem III.2 to conclude that the invariant manifolds defined there are $\psi_s(t')$ invariant for all t' near 0, and, hence, that V_s is tangent to them and that they are $\psi_s(t')$ invariant for all $t' \in \mathbb{R}$.

(d) State the local theorem which follows from these considerations.

EXERCISE III.4. Let $T: E \to E$ be a hyperbolic automorphism of the Banach space E. There is an $s > 0$ and a neighborhood U of T in $C^r(E(s), E)$ such that every f in U has an unstable manifold by Theorem 5.2 which is defined as the graph of a C^r function $\varphi_f: E_1(s) \to E_2(s)$. Consider the joint mapping $(f, x) \to (f, f(x))$ defined on $U \times E_1(s) \times E_2(s)$. By Theorem III.8 observe that this map has a C^r center stable manifold defined as the graph of a C^r function on a perhaps smaller neighborhood $V \times E_1(s') \to E_2(s')$ and conclude that the map $(f, x) \to (\varphi_s(x))$ is C^r in x and f on $V \times E_1(s')$.

Commentary

To see the stable and unstable manifolds and to understand how they intersect, not only at one point, but at every point of a hyperbolic set (see Chapter 6) and then to extend these manifolds near a hyperbolic set (see the commentaries on Chapter 7) is the heart of the subject.

In [4.1], Anosov gives a short historical sketch.

I have closely followed [4.2], adding to the proofs some ideas taken from [5.1]. This is in order, among other things, to enable the reader of Chapters 4 and 5 to read [5.1].

Notice that in the C^r Section Theorem 5.18, X can be a quite arbitrary manifold (infinite dimensional, etc.) as long as TX has an inverse in K theory, e.g., TX is admissible in the sense of (5.17).

The proof of the unstable manifold theorem in Appendix II comes from [5.2] and [5.3].

References

[5.1] Hirsch, M. W., Pugh, C. C. and Shub, M., *Invariant Manifolds*, Lecture Notes in Mathematics, No. 583, Springer-Verlag, New York, 1976.

[5.2] Irwin, M. C., On the stable manifold theorem, *Bull. London Math. Soc.* **2** (1970), 196.

[5.3] Irwin, M. C., On the smoothness of the composition map, *Quart. J. Math. Oxford* **23** (1972), 113; *Bull. London Math. Soc.* **2** (1970), 196.

CHAPTER 6
Stable Manifolds for Hyperbolic Sets

Next, we will generalize to more complicated hyperbolic sets, such as the horseshoe or a solenoid, the theory we have developed for a periodic point.

Definition 6.1. Let f be a diffeomorphism of M and x a point in M. We define

$$W_\varepsilon^s(x, f) = \{y \in M \mid d(f^n(x), f^n(y)) \to 0 \text{ as } n \to +\infty \text{ and } d(f^n(x), f^n(y)) \leq \varepsilon, \forall n \geq 0\},$$

$$W^s(x, f) = \bigcup_{n \geq 0} f^{-n} W_\varepsilon^s(f^n(x)),$$

$$W_\varepsilon^u(x, f) = \{y \in M \mid d(f^n(x), f^n(y)) \to 0 \text{ as } n \to -\infty \text{ and } d(f^n(x), f^n(y)) \leq \varepsilon, \forall n \leq 0\},$$

$$W^u(x, f) = \bigcup_{n \geq 0} f^n W_\varepsilon^u(f^{-n}(x)).$$

Note that this definition is equivalent to the one given in Chapter 2.

Theorem 6.2. *Let Λ be a closed hyperbolic set for f, and assume Λ is furnished with an adapted metric. Then there is a positive ε such that for every point x in Λ, $W_\varepsilon^s(x, f)$ is an embedded disk of dimension equal to that of E_x^s; moreover, $T_x W_\varepsilon^s(x) = E_x^s$; and similarly for the unstable case.*

The stable and unstable discs also satisfy the following:

(1)
$$d(f^n(x), f^n(y)) \leq \lambda^n d(x, y), \forall y \in W_\varepsilon^s(x), \forall n \geq 0,$$
$$d(f^{-n}(x), f^{-n}(y)) \leq \lambda^n d(x, y), \forall y \in W_\varepsilon^u(x), \forall n \geq 0,$$

where $\lambda < 1$ is such that $\|Df|_{E^s}\| < \lambda$ and $\|(Df|_{E^u})^{-1}\| < \lambda$.

(2) *The embedding of* $W_\varepsilon^{u(\text{resp. } s)}(x, f)$ *varies continuously with* x. *More precisely, if* f *is* C^r *and* $n = \dim E^s$, *there is a neighborhood* U *of* x *and a continuous map*

$$\Theta: U \to \text{Emb}^r(D^n, M)$$

such that $\Theta(y)(0) = y$ *and* $\Theta(y)(D^n) = W_\varepsilon^s(y, f)$, $\forall y \in U$.

(3) $$W_\varepsilon^s(x, f) = \{y \mid d(f^n(x), f^n(y)) \leq \varepsilon, \forall n \geq 0\}$$

$$W_\varepsilon^u(x, f) = \{y \mid d(f^n(x), f^n(y)) = \leq \varepsilon, \forall n \leq 0\}.$$

(4) *The manifold* $W_\varepsilon^{u(\text{resp. } s)}(x, f)$ *is as smooth as* f.

We will reduce the proof to an application of the previous stable manifold theorem in a conveniently chosen Banach space. We begin with some preliminaries about the infinite dimensional spaces that we will use.

Let Λ be closed in M and E a vector bundle over Λ. We denote by $\Gamma^0(\Lambda, E)$ the space of continuous sections of E and by $\Gamma^b(\Lambda, E)$ the space of bounded sections. These two spaces have natural vector bundle structures. The space $\Gamma^0(\Lambda, E)$ is contained in $\Gamma^b(\Lambda, E)$ since M and hence Λ are compact. The sup norm on $\Gamma^b(\Lambda, E)$ is defined by $\|h\| = \sup_{x \in \Lambda} \|h(x)\|$. This norm makes Γ^b a Banach space. The set Γ^0 is closed in Γ^b, since the uniform limit of continuous functions is continuous.

Let f be a homeomorphism of M and Λ a closed f-invariant set. We define the automorphism $f_\#$ of $\Gamma^b(\Lambda, T_\Lambda M)$ by

$$f_\#: h \mapsto Df \cdot h \cdot f^{-1},$$

that is, $(f_\#(h))(x) = Df_{f^{-1}(x)}(h(f^{-1}(x)))$ (Figure 6.1). If f is C^1, the automorphism Df of TM is C^0; so $f_\#$ sends $\Gamma^0(\Lambda, T_\Lambda M)$ into itself.

Note that $f_\#$ is a continuous map of $\Gamma^b(\Lambda, T_\Lambda M)$ into itself and $\|f_\#\| = \|Df\|$.

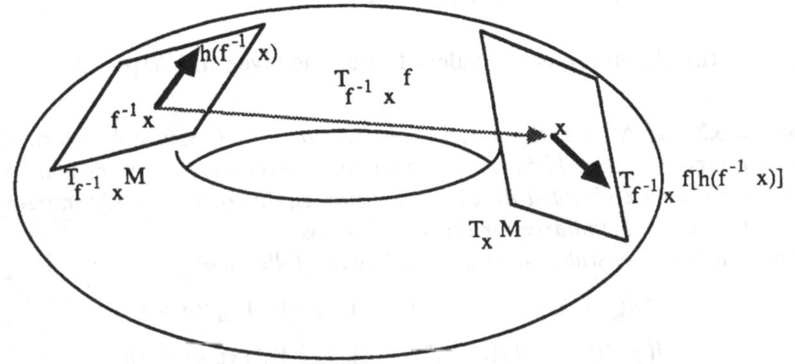

Figure 6.1.

Lemma 6.3. *Let Λ be a closed invariant set for a C^1 diffeomorphism f of M. Λ is a hyperbolic set for f if and only if $f_\#$ is a hyperbolic linear map.*

PROOF. If Λ is a hyperbolic set, then the vector space $\Gamma^b(\Lambda, T_\Lambda M)$ admits a hyperbolic splitting for $f_\#$:

$$\Gamma^b(\Lambda, T_\Lambda M) = \Gamma^b(\Lambda, E^s) \oplus \Gamma^b(\Lambda, E^u).$$

On the other hand, if $\Gamma^b(\Lambda, T_\Lambda M) = E^s \oplus E^u$ is a hyperbolic splitting for $f_\#$, one can reconstruct a hyperbolic splitting of $T_\Lambda M$ as follows:

$$E_x^s = E^s(x) = \{g(x) | g \in E^s\}$$

and similarly for E^u. The details are left to the reader. □

Let $B(\Lambda, M)$ be the space of bounded, not necessarily continuous mappings of Λ into M. The space $B(\Lambda, M)$ is a Banachable manifold modeled on the space of bounded vector fields (sections of the tangent bundle) $\Gamma^b(\Lambda, TM)$. Recall the definitions of the exponential map $\exp_x$ and exponential charts on $B(\Lambda, M)$.

The map $\exp_x$ is the mapping of $T_x M$ to M which is:

(1) tangent to the identity at the origin of $T_x M$,
(2) sends lines in $T_x M$ through the origin to geodesics in M through x,
(3) sends balls in $T_x M$ about the origin to balls in M about x,
(4) $d(\exp_x x_i, x) = \|x_i\|$ for sufficiently small x_i in $T_x M$.

In (3) more is true; for small enough δ, $\exp_x$ is a surjective diffeomorphism of $B(0, \delta)$ onto $B(x, \delta)$ and when M is compact, δ does not depend on x. In this case, the exponential map is a diffeomorphism of a neighborhood of the zero section of the tangent bundle onto a neighborhood of the diagonal in $M \times M$:

$$\exp\colon u \mapsto (m, \exp_m(u)), \qquad \text{where} \quad u \in T_m(M).$$

Charts on $B(\Lambda, M)$. Let inc(Λ) be the inclusion of Λ in M. Let U_δ be the neighborhood of inc(Λ) in $B(\Lambda, M)$, consisting of mappings of satisfying $d(g(x), x) \le \delta$ for all x in Λ. The chart Φ is defined by

$$\Phi\colon U_\delta \to \Gamma_\delta^b(\Lambda, TM) \subset \Gamma^b(\Lambda, TM),$$

$$\Phi\colon h \to \exp^{-1}(\mathrm{graph}(h)).$$

$\Phi(h)$ is thus the section given by $\Phi(h)(x) = \exp_x^{-1}(h(x))$, that is, $\Phi(h)(x) = \exp^{-1}(x, h(x))$.

Note that $\Phi(\mathrm{inc}(\Lambda))$ is the zero section of TM which we denote by $\tilde{0}$.

The mappings $\hat{F}$ *and* $\tilde{F}$. We associate to f the automorphism $\hat{F}$ of $B(\Lambda, M)$ defined by $\hat{F}(h) = fhf^{-1}$. The inclusion inc(Λ) is thus a fixed point of $\hat{F}$. We will show that inc(Λ) is, in fact, a hyperbolic fixed point of $\hat{F}$. By using exponential charts, it suffices to study the mapping $\tilde{F}$ defined on a neighborhood of the zero section by $\tilde{F} = \Phi \hat{F} \Phi^{-1}$:

$$\Gamma^b_\eta(\Lambda, TM) \xrightarrow{\Phi^{-1}} U(\text{inc}(\Lambda)) \xrightarrow{\hat{F}} B(\Lambda, M) \xrightarrow{\Phi} \Gamma^b_\eta(\Lambda, TM).$$

The image of the section σ of $\Gamma^b_\eta(\Lambda, TM)$, for small $\eta < \delta$, is the section

$$\tilde{F}(\sigma)(x) = \exp_x^{-1}(f(\exp_{f^{-1}x}(\sigma(f^{-1}(x))))).$$

The mapping $\tilde{F}$ is as smooth as f and has for its derivative at $\tilde{0}$ the automorphism $f_\# = D_{\tilde{0}}\tilde{F}$ of $\Gamma^b(\Lambda, TM)$ defined by

$$f_\#(\sigma) = Df \cdot \sigma \cdot f^{-1}.$$

Lemma 6.3 allows us to see that $f_\#$ is linear and hyperbolic, so that the zero section $\tilde{0}$ is a hyperbolic fixed point for $\tilde{F}$; that is to say, inc(Λ) is a hyperbolic fixed point for $\hat{F}$. The bundle $\Gamma^b(\Lambda, TM)$ has the following hyperbolic splitting for $f_\# = D_{\tilde{0}}\tilde{F}$:

$$\Gamma^b(\Lambda, TM) = \Gamma^b(\Lambda, E^s) \oplus \Gamma^b(\Lambda, E^u),$$

where $T_\Lambda M = E^s \oplus E^u$ is a hyperbolic splitting of TM for f.

The stable manifold of $\tilde{0}$. Applying the stable manifold theorem to $\tilde{0}$, we see that there is an invariant manifold for $\tilde{F}$: $\tilde{W}^s_\eta(\tilde{0}, \tilde{F})$ which is the graph of a C^r function:

$$\psi\colon \Gamma^b_\eta(\Lambda, E^s) \to \Gamma^b_\eta(\Lambda, E^u).$$

The stable manifold $\tilde{W}_\eta$ which we have found is defined relative to the box norm $\| \ \|'$ on $\Gamma^b(\Lambda, TM) = \Gamma^b(\Lambda, E^s) \oplus \Gamma^b(\Lambda, E^u)$ defined by $\|(x^s, x^u)\|' = \max(\|x^s\|, \|x^u\|)$.

The stable manifold $\tilde{W}^s_\eta(\tilde{0}, \tilde{F})$ satisfies

$$\tilde{W}^s_\eta(\tilde{0}, \tilde{F}) = \{\sigma \in \Gamma^b_\eta(\Lambda, TM) | \tilde{F}^n(\sigma) \in \Gamma^b_\eta(\Lambda, TM), \forall n \geq 0\},$$

where if $\tau \in \Gamma^b_\eta(\Lambda, E^s)$, $\psi(\tau) \in \Gamma^b_\eta(\Lambda, E^u)$ is the only section of the unstable bundle which satisfies

$$\forall n \geq 0, \qquad (\tilde{F})^n(\tau, \psi(\tau)) \in B'_\eta(\tilde{0}),$$

where $B'_\eta(\tilde{0})$ is the η ball about $\tilde{0}$ in $\Gamma^b(\Lambda, TM)$ with the box norm.

Returning to the norm $\| \ \|$ on $\Gamma^b(\Lambda, TM)$ induced by the Riemannian metric on M, we set

$$W^s_\varepsilon(\tilde{0}, \tilde{F}) = \{\sigma \in \Gamma^b(\Lambda, TM) | \|\tilde{F}^n(\sigma)\| \leq \varepsilon, \forall n \geq 0\}.$$

Since the norms $\| \ \|$ and $\| \ \|'$ are equivalent we have, for small enough ε,

$$W^s_\varepsilon(\tilde{0}, \tilde{F}) \subset \tilde{W}^s_\eta(\tilde{0}, \tilde{F}).$$

Lemma 6.4. $W^s_\varepsilon(x, f) = \{\exp_x(\sigma(x)) | \sigma \in W^s_\varepsilon(\tilde{0}, \tilde{F})\} = \{h(x) | h \in W^s_\varepsilon(\text{inc}(\Lambda), \hat{F})\}$
$[W^s_\varepsilon(\tilde{0}, \tilde{F}) \subset \Gamma^b(\Lambda, TM);\ W^s_\varepsilon(\text{inc}(\Lambda), \hat{F}) \subset B(\Lambda, M)]$.

PROOF. Let h be a point in the stable manifold $W_\varepsilon^s(\mathrm{inc}(\Lambda), \hat{F})$; the sequence $\hat{F}^n(h)$ tends to $\mathrm{inc}(\Lambda)$, thus

$$\sup_{z\in\Lambda} d[\hat{F}^n(h)(z),\ \hat{F}^n(\mathrm{inc}(\Lambda))(z)] \to 0 \qquad \text{as} \quad n \to +\infty.$$

Therefore, since $\mathrm{inc}(\Lambda)$ is fixed by $\hat{F}$, we have

$$\sup_{z\in\Lambda} d[\hat{F}^n(h)(z), z] \to 0 \qquad \text{as} \quad n \to +\infty,$$

that is

$$\sup_{z\in\Lambda} d[f^n h f^{-n}(z), z] \to 0 \qquad \text{as} \quad n \to +\infty.$$

Thus

$$\sup_{z\in\Lambda} [f^n h(z), f^n z] \to 0 \qquad \text{as} \quad n \to +\infty$$

and so $h(z) \in W_\varepsilon^s(z, f)$.

Conversly, supposing that y is in $W_\varepsilon^s(x, f)$, we have $d(f^n(x), f^n(y)) \to 0$ as $n \to +\infty$. Defining a bounded function $\delta_x^y \in B(\Lambda, M)$ by

$$\delta_x^y(x) = y,$$

$$\delta_x^y(z) = z, \qquad \forall z \neq x,$$

we see

$$\hat{F}(\delta_x^y(z)) = f \circ \delta_x^y \circ f^{-1}(z) = \begin{cases} z & \text{if } z \neq f(x), \\ f(y) & \text{if } z = f(x), \end{cases}$$

that is, $\hat{F}(\delta_x^y) = \delta_{f(x)}^{f(y)}$, so $\hat{F}^n(\delta_x^y) = \delta_{f^n(x)}^{f^n(y)}$.

We will now show that if y belongs to $W_\varepsilon^s(x, f)$ the map δ_x^y belongs to $W_\varepsilon^s(\mathrm{inc}(\Lambda), \hat{F})$. In fact

$$\begin{aligned} d[\hat{F}^n(\delta_x^y), \mathrm{inc}(\Lambda)] &= \sup_{z\in\Lambda} d[\hat{F}^n(\delta_x^y)(z), \mathrm{inc}(\Lambda)(z)] \\ &= \sup_{z\in\Lambda} d[\delta_{f^n(x)}^{f^n(y)}(z), z] \\ &= d[f^n(y), f^n(x)]. \end{aligned}$$

Now $d[f^n(y), f^n(x)]$ tends to 0 as n tends to ∞, and we are done; δ_x^y does belong to $W_\varepsilon^s(\mathrm{inc}(\Lambda), \tilde{F})$ for y in $W_\varepsilon^s(x, f)$.

We have then

$$W_\varepsilon^s(x, f) = \{h(x) \mid h \in W_\varepsilon^s(\mathrm{inc}(\Lambda), \hat{F})\}$$

or, using exponential charts,

$$W_\varepsilon^s(x, f) = \{\exp_x \gamma(x) \mid \gamma \in W_\varepsilon^s(\tilde{0}, \tilde{F})\}. \qquad \square$$

Lemma 6.5. *There is a continuous bundle map μ, C^r on each fiber such that the image of a section σ in $\Gamma_\eta^b(\Lambda, E^s)$ under the mapping ψ which defines $\tilde{W}_\eta^s(\tilde{0}, \tilde{F})$ can be written $\psi(\sigma) = \mu \circ \sigma$.*

The diagram

$$\begin{array}{ccc} E_\eta^{\mathrm{s}} & \xrightarrow{\mu} & E_\eta^{\mathrm{u}} \\ \big\downarrow & & \big\downarrow \\ \Lambda & \xrightarrow{\mathrm{id}} & \Lambda \end{array}$$

commutes.

Moreover, the restriction of μ to the fiber over x along with its derivatives of order up to r, depends continuously on x.

Proof. The condition that $\psi(\sigma) = \mu \circ \sigma$ implies in particular that the value $\psi(\sigma)$ at x depends only on $\sigma(x)$ and not on the values of σ elsewhere.

Let, then, x be a point in Λ, and σ_1 and σ_2 be two sections in $\Gamma_\eta^{\mathrm{b}}(\Lambda, E^{\mathrm{s}})$ having the same value at x: $\sigma_1(x) = \sigma_2(x)$; we will show that $\psi(\sigma_1)(x) = \psi(\sigma_2)(x)$.

Suppose that $\psi(\sigma_1)(x) \neq \psi(\sigma_2)(x)$. However, we know that $\tilde{F}^n(\sigma_1, \psi(\sigma_1)) \to \tilde{0}$ and $\tilde{F}^n(\sigma_2, \psi(\sigma_2)) \to \tilde{0}$, as $n \to +\infty$, since $[\sigma_1, \psi(\sigma_1)]$ and $[\sigma_2, \psi(\sigma_2)]$ belong to $\tilde{W}_\eta^{\mathrm{s}}(\tilde{0}, \tilde{F})$.

Now define a bounded map, $\tau \in \Gamma_\eta^{\mathrm{b}}(\Lambda, E^{\mathrm{u}})$, by

$$\tau(y) = \psi(\sigma_1)(y) \qquad \text{for} \quad y \neq x,$$
$$\tau(x) = \psi(\sigma_2)(x).$$

From above, we know that $\tilde{F}^n(\sigma_1, \tau) \to \tilde{0}$ when $n \to +\infty$, which is absurd if $\psi(\sigma_1)(x) \neq \psi(\sigma_2)(x)$, because it contradicts the fact that ψ could also be defined by

$$\psi(\sigma_1) \text{ is the only section of } E_\eta^{\mathrm{u}} \text{ such that } \tilde{F}^n(\sigma_1, \psi(\sigma_1)) \to \tilde{0}.$$

We have, in fact

$$\begin{aligned} d[\tilde{F}^n(\sigma_1, \tau), \tilde{0}] &= \sup_{z \in \Lambda} d[\tilde{F}^n(\sigma_1, \tau)(z), 0_z] \\ &= \sup_{z \in \Lambda} \|\exp_z^{-1}[f^n(\exp_{f^{-n}(z)}(\sigma_1, \tau)(f^{-n}(z)))]\| \\ &= \max\Bigg\{ \sup_{\substack{z \in \Lambda \\ z \neq f^n(x)}} \|\exp_z^{-1}[f^n(\exp_{f^{-n}(z)}(\sigma_1, \psi(\sigma_1))(f^{-n}(z)))], \\ &\qquad \|\exp_{f^n(x)}^{-1}[f^n(\exp_x(\sigma_2, \psi(\sigma_2))(x))]\| \Bigg\} \\ &\leq \max\{d[\tilde{F}^n(\sigma_1, \psi(\sigma_1)), \tilde{0}], d[\tilde{F}^n(\sigma_2, \psi(\sigma_2)), \tilde{0}]\}. \end{aligned}$$

The last expression tends to 0, when n tends to $+\infty$, which shows that replacing $\psi(\sigma_1)(x)$ by $\psi(\sigma_2)(x)$ does not keep $\tilde{F}^n(\sigma_1, \tau)$ from converging uniformly to the zero section. Thus, $\psi(\sigma_1)(x) = \psi(\sigma_2)(x)$, which shows that $\psi(\sigma)(x)$ only depends on $\sigma(x)$, as we hoped.

We can now define the mapping $\mu\colon E^s_\eta \to E^u_\eta$ by $\mu(v) = \psi(\delta^v_x)(x)$, recalling that δ^v_x is the section given by $\delta^v_x(x) = v$, $\delta^v_x(z) = 0_z$ for $z \neq x$. In short, μ is defined by the following composition:

$$\begin{array}{ccccccc} E^s_x & \to & \Gamma^b_\eta(\Lambda, E^s) & \xrightarrow{\psi} & \Gamma^b_\eta(\Lambda, E^u) & \xrightarrow{ev_x} & E^u_x, \\ v & \mapsto & \delta^v_x & \mapsto & \psi(\delta^v_x) & \mapsto & \psi(\delta^v_x)(x) = \mu(v). \end{array}$$

This definition guarantees that μ preserves the fibers of E^s_η. Furthermore, μ is induced by the mapping $\psi\colon \Gamma^b_\eta(\Lambda, E^s) \to \Gamma^b_\eta(\Lambda, E^u)$, which is C^r; thus we easily see that the restriction of μ to the fiber of E^s_η over x, μ_x, is a C^r map.

It only remains to show that μ_x depends continuously on x in the C^r topology, which is not immediate since we have considered bounded sections without regard for the topology on Λ.

To prove the continuity, it suffices to show that μ induces a C^r map of the space of continuous sections $\Gamma^0_\eta(\Lambda, E^u)$.

Recall that we have the following commutative diagram:

$$\begin{array}{ccc} C^0(\Lambda, M) & \xrightarrow{\hat{F}} & C^0(\Lambda, M) \\ \text{inclusion}\downarrow & & \downarrow\text{inclusion} \\ B(\Lambda, M) & \xrightarrow{\hat{F}} & B(\Lambda, M). \end{array}$$

The inclusion $\mathrm{inc}(\Lambda)$ is a hyperbolic fixed point for the mapping $\hat{F}\colon C^0(\Lambda, M) \to C^0(\Lambda, M)$. Thus we can, working with charts, produce the stable manifold of $\mathrm{inc}(\Lambda)$ as the graph of a map

$$\psi'\colon \Gamma^0_\eta(\Lambda, E^s) \to \Gamma^0_\eta(\Lambda, E^u)$$

This map ψ' is C^r and must be the restriction of ψ to $\Gamma^0_\eta(\Lambda, E^s)$, since the stable manifold of 0 in $\Gamma^0_\eta(\Lambda, TM)$ is the graph of ψ. Thus $\psi'(\sigma) = \mu \circ \sigma$, and we see the maps μ_x and their derivatives up to order r depend continuously on x. □

End of the Proof of Theorem 6.2. We have

$$W^s_\varepsilon(x, b) = \{\exp_x(\gamma(x)) \mid \gamma \in W^s_\varepsilon(\tilde{0}, \tilde{F})\} \subset \tilde{W}^s_\eta(x, b),$$

where

$$\tilde{W}^s_\eta(x, b) = \{\exp_x[v, \mu(v)] \mid v \in E^s_x, \|v\| < \eta\}$$

or, equally,

$$\tilde{W}^s_\eta(x, b) = \{\exp_x(\mathrm{graph}\, \mu_x \subset T_x M)\}.$$

Since the restriction of μ to each fiber is C^r and since $\exp_x$ is a local diffeomorphism, $\tilde{W}^s_\eta(x, b)$ is an embedded disc through x of dimension equal to that of E^s_x (Figure 6.2). Since $\psi(\tilde{0}) = \tilde{0}$, we have $\mu \circ \tilde{0} = \tilde{\psi}(\tilde{0}) = \tilde{0}$, so $\mu(0_{E^s_x}) = 0_{E^u_x}$. Since the graph of ψ, $\{(\sigma, \psi(\sigma)) \mid \sigma \in \Gamma^b_\varepsilon(\Lambda, E^s)\}$, is tangent to $\Gamma^b(\Lambda, E^s)$ at $\tilde{0}$, the graph of the restriction of μ to the fiber over x, $\{(v, \mu(v)) \mid v \in E^s_x, \|v\| < \eta\}$ is tangent to E^s_x at the origin of the tangent space to $\mathbf{M}$ at x. Therefore

$$T_x(\tilde{W}^s_\eta(x, b)) = E^s_x.$$

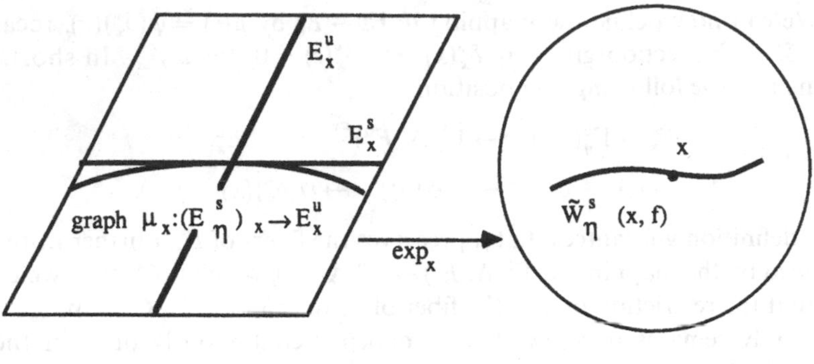

Figure 6.2.

Consider the map

$$(\text{graph } \mu): \Lambda \to \text{Emb}(D^s, M),$$
$$x \mapsto \exp_x[\text{graph } \mu_x],$$

where $s = \dim E^s_x$. This map is defined locally by means of charts for the bundle E^s and is continuous.

Let us decompose this map. Let C_x: $D^s_\eta \to (E^s_\eta)_x$ be a trivializing chart for the bundle in a neighborhood of x. Let p_x: $D^s_\eta \to M$ be defined by $p_x(y) = \exp_x[C_x(y), \mu \circ C_x(y)]$. Purists should not forget to rescale via sc: $D^s_1 \to D^s_\eta$. The map (graph μ) is defined locally by (graph $\mu)x = p_x \circ sc$, and is clearly continuous.

Consequently, $\tilde{W}^s_\eta(x, f)$ is a C^r disc which depends continuously on x in the C^r topology.

We now show that $W^s_\varepsilon(x, f)$ is a disc of the same dimension as $\tilde{W}^s_\eta(x, f)$ and also depends on C^r continuously on x.

First notice that the norm $\|T_x f|_{\tilde{W}^s_\eta(x,f)}\|$ is strictly less than λ, since the tangent space to $\tilde{W}^s_\eta(x, f)$ is E^s_x. Thus we have, for sufficiently small ε, the following implication: if y is in $\tilde{W}^s_\eta(x, f)$ and $d(x, y) \leq \varepsilon$, then $d(f(x), f(y)) \leq \lambda d(x, y)$. (One can choose ε independent of x since Λ is compact and $\tilde{W}^s_\eta(x, f)$ depends continuously on x.)

Thus we have

$$W^s_\varepsilon(x, f) = \{y \in \tilde{W}^s_\eta(x, f) | d(y, x) \leq \varepsilon\} = \tilde{W}^s_\eta(x, b) \cap B_\varepsilon(x).$$

Now, if ε is very small, $\tilde{W}^s_\eta(x, b) \cap B_\varepsilon(x)$ is a disc and Thom's isotopy theorem shows that it depends C^r continuously on x. □

Corollary and Definition 6.6. *Under the hypothesis of* 6.2, *if* $x \in \Lambda$, *the set* $W^s(x, f) = \bigcup_{n \geq 0} f^{-n}[W^s_\varepsilon(x, f^n x)]$ *is an immersed submanifold of* M. *We call this submanifold the global stable manifold of* x *for* f *in contrast to the local stable manifold* $W^s_\varepsilon(x, f)$. *Of course, there are analogous definitions for the unstable case.*

Appendix IV

Center and Strong Stable Manifolds for Invariant Sets

The stable manifold theorem for hyperbolic sets may be generalized to strong stable, center stable, center, center unstable, and strong unstable manifolds for hyperbolic invariant sets using Theorem III.2 or 8. Here we will only state the strong stable and center stable manifold theorem. The strong unstable and center unstable theorems are proven by replace f by f^{-1} and the center manifolds are obtained by intersecting the center unstable and center stable.

Theorem IV.1. *Let Λ be an invariant set for the C^r diffeomorphism f of M. Suppose that $TM|\Lambda$ has a continuous Tf invariant direct sum decomposition*

$$TM|\Lambda = E_1 \oplus E_2$$

and that there are real constants

$$0 < \lambda < \rho < \mu \quad \text{and} \quad 0 < \lambda < 1$$

such that

$$\|Df(x)v\| \leq \lambda \|v\| \qquad \textit{for all} \quad x \in \Lambda \quad \textit{and} \quad v \neq 0 \textit{ in } E_2$$

and

$$\|Df(x)v\| \geq \mu \|v\| \qquad \textit{for all} \quad x \in \Lambda \quad \textit{and} \quad v \neq 0 \textit{ in } E_1$$

for an adapted metric in $TM|\Lambda$. There is a positive ε and for every point x in Λ two embedded discs $W_\varepsilon^{ss}(x)$ and $W_\varepsilon^{cu}(x)$ tangent at x to $E_2(x)$ and $E_1(x)$, respectively, and satisfying:

(0) *$W_\varepsilon^{ss}(x)$ is a C^r embedded disc.*
(1) *$W_\varepsilon^{ss}(x) = \{y | d(f^n(x), f^n(y)) \leq \varepsilon,\ n \geq 0,$ and $d(f^n(x), f^n(y))/\rho^n \to 0$ as $n \to \infty\}$.*
(2) *$f(W_\varepsilon^{ss}(x)) \subseteq W_\varepsilon^{ss}(f(x))$ and f contracts distance by a constant close to λ.*
(3) *The embedding $W_\varepsilon^{ss}(x)$ varies continuously with x. More precisely, if f is C^r and $n = \dim E_2$ there is a neighborhood U of x and a continuous map Θ: $U \to \mathrm{Emb}^r(D^n, M)$ such that*

$$\Theta(x)(0) = x \quad \textit{and} \quad \Theta(x)(D^n) = W_\varepsilon^{ss}(x).$$

(4) *If $\lambda\rho^{-j} < 1$ for $1 \leq j \leq r$ then $W^{cu}(x)$ is C^r.*
(5) *$f(W_\varepsilon^{cu}(x)) \cap B_\varepsilon(x) \subseteq W_\varepsilon^{cu}(f(x))$, where*

$$B_\varepsilon(x) = \{y \in M | d(x, y) < \varepsilon\}.$$

(6) *The $W_\varepsilon^{cu}(x)$ vary continuously as C^r embedded discs as in* (3).

Proof. The proof of this theorem is almost the same as Theorem 6.2 except that we apply Theorem III.2 or 8 instead of the stable manifold theorem.

$\Gamma^b(\Lambda, T_\Lambda M) = \Gamma^b(\Gamma, E_1) \oplus \Gamma^b(\Lambda, E_2)$ which is a ρ-pseudohyperbolic splitting for the linear map $f_\#: \Gamma^b(\Lambda, T_\Lambda M) \to \Gamma^b(\Lambda, T_\Lambda)$ defined by $f_\#(h) = Df \circ h \circ f^{-1}$. The maps $\hat{F}: B(\Lambda, M) \to B(\Lambda, M)$ defined by $\hat{F}(h) = fhf^{-1}$ and $\tilde{F}$ defined on a neighborhood of the zero section in $\Gamma^b(\Lambda, TM)$ by $\tilde{F}(\sigma)(x) = \exp_x^{-1}(f(\exp_{f^{-1}x}(\sigma(f^{-1}(x)))))$ are as smooth as f, $D_{\bar{0}}\tilde{F}: \Gamma^b(\Lambda, TM) \to \Gamma^b(\Lambda, TM)$ and $D_{\bar{0}}\tilde{F} = f_\#$. In order to apply Theorem III.2 or 8 we produce C^r extensions $\tilde{F}_s$ of $\tilde{F}$ on the ball of radius s in $\Gamma^b(\Lambda, TM)$ such that $\mathrm{Lip}(F_s - f_\#)$ tends to zero as $s \to 0$. The function $g_x = \exp_{f(x)}^{-1} f \exp_x$ maps a neighborhood of zero in T_xM into $T_{f(x)}M$ and has derivative Df_x at 0 in T_xM. Suppose that s is small enough, and $\varphi_{s,x} = \varphi_{1,x}(1/s)$ is a standard bump function defined on T_xM and varying smoothly with x, then

$$g_{s,x}(v) = \begin{cases} \varphi_{s,x}(v)g_x(v) + (1 - \varphi_{s,x}(v))Df_x(0)v & \text{for } x \in B(2s), \\ Df_x(0)v & \text{for } x \text{ in the complement of } B(2s), \end{cases}$$

extends g_x on $B(s)$ to a diffeomorphism $g_{s,x}: T_xM \to T_{f(x)}M$, as smooth as f and $\mathrm{Lip}(g_{s,x} - Df_x(0)) \to 0$ as $s \to 0$, uniformly in x. Now extend $\tilde{F}$ to $\tilde{F}_s$ by $\tilde{F}_s\sigma(x) = g_{s,x}\sigma f^{-1}(x)$. $\tilde{F}_s$ is C^r, extends $\tilde{F}$ on $B(s)$ in $\Gamma^b(\Lambda, TM)$ and $\mathrm{Lip}(\tilde{F}_s - f_\#) \to 0$ with s as the same is true on each fiber, that is,

$$\mathrm{Lip}(\tilde{F}_s - f_\#) \le \sup_{x \in \Lambda} \mathrm{Lip}(g_{s,x} - Df_x(0))$$

which tends to zero. Now Theorem III.2 may be applied to $\tilde{F}_s$. The proof now proceeds as the proof of 6.2, using the characterizations of W^{ss} and W^{cu} which follow from III.2. We leave the construction of the function μ and the rest of the proof as an exercise. □

Note that, of course, the $W^{cu}(x)$ are not necessarily unique, they may depend on the extension of $\tilde{F}$ constructed. The W^{ss} are, however, unique since they are locally dynamically defined.

Exercise IV.1. Prove that the function μ of Theorem IV.1 exists and finish the proof of the theorem.

Exercise IV.2. Formulate center, center unstable, and strong unstable manifold theorems for invariant sets which are the analogue of Theorem III.8.

Commentary

I have retained the presentation of [4.2]. Moser [6.2] gives a new proof of the theorem of Anosov. In the appendix to [1.16], Mather translates the implicit function theorem techniques of Moser in terms of Banach manifolds. The idea is the following: to find a continuous solution of the equation $gh = hf$, when g is close to f, we examine the transformation $T_g: C^0(M, M) \to C^0(M, M)$ defined by $T_g(h) = ghf^{-1}$. When $g = f$, the identity, id_M; is a solution, that is, a fixed point of T_f. It is a transverse fixed point if and only if $I - f_\#$ is invertible,

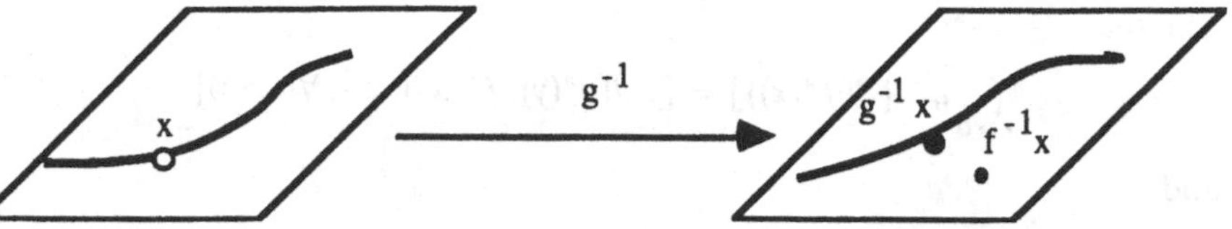

Figure C6.1.

since, in appropriate charts, $f_{\#}$ is the derivative of T_f at id_M. The implicit function theorem implies that the transformation T_g has a unique fixed point in a neighborhood of id_M when g is close enough to f. Mather showed that the endomorphism $I - f_{\#}$ of $\Gamma^0(M, TM)$ is invertible if and only if f is an Anosov diffeomorphism. I mention this here to show, historically, how attention was focused on $I - f_{\#}$ and to expose the roots of the proof. (Figure C6.1).

Note that the fixed point of T_g is the intersection (in $C^0(M, M)$) of the local stable and unstable manifolds of T_g. Once again, if we follow the proof of the stable and unstable manifold theorems, we conclude that, if g is close enough to f,

$$\exp[ev_x W^u_{loc}(T_g)] = \exp(\mathrm{graph}(U_x))$$

is the image under the exponential map of the graph of a function $U_x: E^u_{x,\varepsilon} \to E^s_{x,\varepsilon}$, that if f and g are sufficiently smooth, then U_x will be C^r, etc. For $\varepsilon > 0$ small enough, $\exp(\mathrm{graph}(U_x))$ will be a C^r manifold and will be composed of points y satisfying $d(g^n(y), f^n(x)) < \varepsilon$, $\forall n \leq 0$. The same analysis applies to the stable sets.

The point of intersection of these two submanifolds is unique: it is $h(x)$ and the two manifolds are transverse here (Figure C6.2). Now choose a box $S(x)$ around each point x of M, and now let g be a map close to f (Figure C6.3).

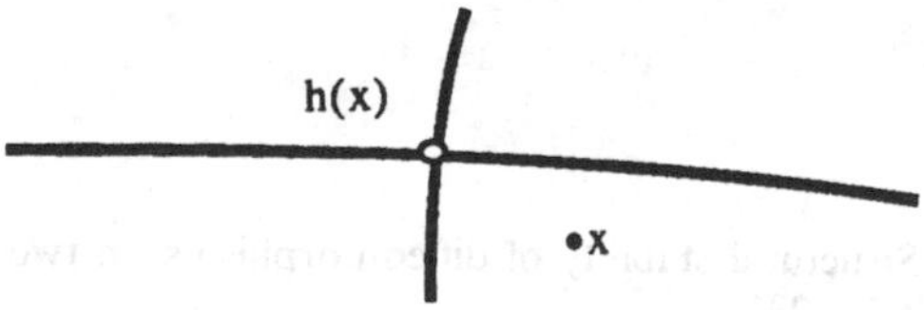

Figure C6.2.

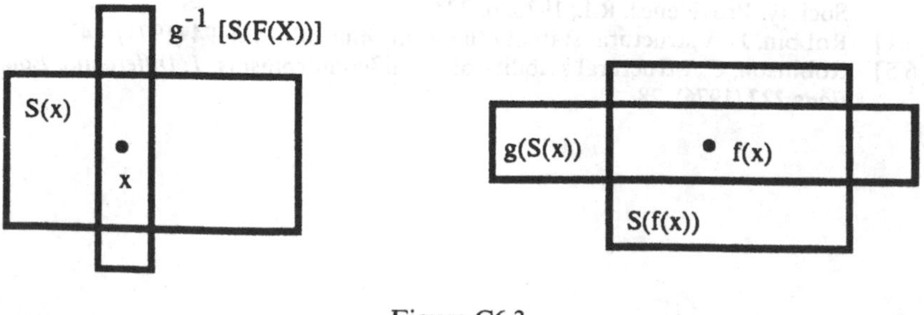

Figure C6.3.

The sets

$$\bigcap_{n\geq 0} g^{-n}[S(f^n(x))] = \{y \mid d(g^n(y), f^n(x)) < \varepsilon, \forall n \geq 0\}$$

and

$$\bigcap_{n\geq 0} g^{n}[S(f^{-n}(x))] = \{y \mid d(g^n(y), f^n(x)) < \varepsilon, \forall n \leq 0\}$$

are manifolds which coincide with the stable and unstable manifolds for f of x, when $g = f$; the map g contracts and expands these local manifolds by a factor close to the constant of hyperbolicity of f, which shows that they are the stable and unstable manifolds for g; finally they are transverse and thus the intersection $\bigcap_{n\in\mathbb{Z}} g^{-n}[S(f^n(x))]$ reduces to a point y which is the only point satisfying $d(g^n(y), f^n(x)) < \varepsilon$, for all n in $\mathbb{Z}$. The sought for conjugacy h is therefore given by $h(x) = y$. This is how Anosov constructed his conjugacy.

One can find the manifolds $\bigcap_{n\geq 0} g^{-n}[S(f^n(x))]$ and $\bigcap_{n\geq 0} g^{n}[S(f^{-n}(x))]$ directly with the aid of a graph transformation on the disjoint union of the boxes $S(x)$.

In general, one can either argue in function space or directly on the manifold M; the latter approach has the advantage of a more geometric flavor, but the precise results are difficult to establish. I have, in general, chosen the functional approach because it seems quicker.

The reader will find these different points of view in Anosov [4.1], Palis and Smale [6.3], Melo [6.1], Robinson [6.5], and Conley [1.5], on the geometric side, and in Moser [6.2] and Robbin [6.4] a more functional approach.

The discussion will continue in the subsequent commentaries.

Appendix IV is taken from [5.1] where there is additional discussion of these points.

References

[6.1] Melo, W., Structural stability of diffeomorphisms on two manifolds, *Invent. Math.* **21** (1973), 233.

[6.2] Moser, J., On a theorem of Anosov, *J. Differential Equations* **5** (1969), 411.

[6.3] Palis, J. and Smale, S., Structural stability theorems, in *Global Analysis*, Vol. XIV (Proceedings of Symposia in Pure Mathematics), American Mathematical Society, Providence, R.I., 1970, p. 223.

[6.4] Robbin, J., A structural stability theorem, *Ann. of Math.* **94** (1971), 447.

[6.5] Robinson, C., Structural stability of C^1 diffeomorphisms, *J. Differential Equations* **222** (1976), 28.

CHAPTER 7

More Consequences of Hyperbolicity

Definition 7.1. Consider two submanifolds V and W of M which intersect at a point p. We say that V and W are *transverse* at p, $V \pitchfork W$, or that p is a *point of transverse intersection* of V and W, if

$$T_0 V + T_0 W = T_p M.$$

More generally, if f is a smooth map of a manifold V into M, and W is a submanifold of M, we say that f is transverse to W at a point p of V, $f \pitchfork W$, if either $f(p) \notin W$ or $f(p) \in W$ and $Df_p(T_p V) + T_{f(p)} W = T_{f(p)} M$. We say that V and W are transverse, $V \pitchfork W$, if V and W are transverse wherever they intersect; and that $f\colon V \to M$ is transverse to W on a subset K if f is transverse to W at all points of K.

When we have a map f from a manifold V of dimension j to a manifold M of dimension m, transverse at p to a submanifold W of dimension k with $f(p)$ in W, there is a chart (U, Φ) on M is a neighborhood of $f(p)$ such that:

(1) $\Phi(W) = \mathbb{R}^k \times \{0\} \subset \mathbb{R}^m$.
(2) The composition $P_i \circ \Phi \circ f$, where P_i is the projection onto the complementary $\mathbb{R}^{m-k}$ of $\Phi(W)$, is defined in a neighborhood of p in V and has a surjective derivative at p.

The implicit function theorem allows us to conclude that, in a neighborhood of p, $[P_i \circ \Phi \circ f]^{-1}(0) = f^{-1}(W)$ is a submanifold of dimension $\dim V + \dim W - \dim M$ of V, in other words, $\operatorname{codim}_V f^{-1}(W) = \operatorname{codim}_M(W)$. When V is compact and W is closed in M, the preimage $f^{-1}(W)$ is a submanifold of V whenever f is transverse to W. Furthermore, transversality is stable, if V and W have no boundaries, any map g, sufficiently C^1 close to an f with $f \pitchfork W$ is also transverse to W and $g^{-1}(W)$ is diffeomorphic to $f^{-1}(W)$. If ∂_V and ∂_W

are nonempty, the preceding still holds, as long as there is a neighborhood of $f(\partial_V) \cup \partial_W$ disjoint from $f(V) \cap W$; this is the case for two discs of complementary dimension intersecting transversely in a single point in their respective interiors, in this case the preceding remarks are equivalent to the local inverse function theorem.

Proposition 7.2. *Let Λ be a compact hyperbolic invariant set for a C^r diffeomorphism of M, $r \geq 1$. Then, for every small positive η there is a positive δ, such that $\forall x, y \in \Lambda$ such that $d(x, y) < \delta$, $W_\eta^s(x) \cap W_\eta^u(y) = p$, where p is a point of transverse intersection of $W_\eta^s(x)$ and $W_\eta^u(y)$.*

Proof. Choose ε to be as in the proof of the stable manifold theorem for Λ. Then $W_\varepsilon^s(x) \cap W_\varepsilon^u(x)$ is a single point x and, in fact, $W_\varepsilon^s \pitchfork W_\varepsilon^u$ since $T_x W_\varepsilon^s(x) = E_x^s$, $T_x W_\varepsilon^u(x) = E_x^u$ and $T_x M = E_x^s \oplus E_x^u$. The proposition then follows from the preceding remarks and the continuity $W_\varepsilon^u(x)$ and $W_\varepsilon^s(x)$ in the C^1 topology. □

Definition 7.3. Let X be a metric space and f a homeomorphism of X to itself. We say that f is *expansive* on a subset Y of X, if there is positive ε such that for any pair of distinct points x and y, x in X and y in Y:

$$\sup_n d[f^n(x), f^n(y)] > \varepsilon.$$

When $X = Y$ we simply say f is expansive.

Proposition 7.4. *Let Λ be a closed hyperbolic invariant set for a C^r diffeomorphism f of M, $r \geq 1$; then f is expansive on Λ.*

Proof. With ε as in the stable manifold theorem, if we have $\sup_n d[f^n(x), f^n(y)] \leq \varepsilon$ for x in M and y in Λ, then we must have $x \in W_\varepsilon^s(y)$ and $x \in W_\varepsilon^u(y)$. Thus, $x \in W_\varepsilon^s(y) \cap W_\varepsilon^u(y)$, so $x = y$. □

Hyperbolic invariant sets are very robust. We will begin by discussing the linear case, using our old friend the graph transform. First, we will show that the hyperbolic linear automorphisms of a Banach space form an open set.

Proposition 7.5. *Given three real numbers, τ, ε, and k, $0 < \tau < 1$, $\varepsilon > 0$, $k > 0$; there is positive δ such that if E is a Banach space, T a hyperbolic automorphism of E with adapted splitting $E = E_1 \oplus E_2$, L a continuous linear automorphism, and we have*

$$\|T^{-1}\| < k, \qquad \|T^{-1}|_{E_1}\| \leq \tau, \qquad \|T|_{E_2}\| \leq \tau, \qquad \|T - L\| < \delta,$$

then L is also hyperbolic with respect to a splitting $E = F_1 \oplus F_2$ and, further,

$$\|L^{-1}|_{F_1}\| < \tau + \varepsilon, \qquad \|L|_{F_2}\| < \tau + \varepsilon.$$

Note that the estimates made below in order to prove the proposition are taken relative to the box norm on E defined by $\| \ \|_E = \max[\| \ \|_{E_1}, \| \ \|_{E_2}]$.

Proof. Let $\mathbf{L}_1(E_1, E_2)$ be the space of continuous linear maps of E_1 to E_2 of norm less than or equal to 1. The graph transform $\Gamma_{\mathbf{L}}: \mathbf{L}_1(E_1, E_2) \to \mathbf{L}_1(E_1, E_2)$ associated to L is well defined when $\|T - L\| < 1 - \tau$, and has a fixed point σ when $\|T - L\| < (1 - \tau)/2$. The mapping σ then gives rise to the unstable manifold, graph (σ), which is a linear subspace F_1 of E. We have, by Lemma 5.5,

$$\|L^{-1}|_{F_1}\| < \frac{1}{1/\tau - \|T - L\|}$$

and thus if $1/(1/\tau - \delta) < \tau + \varepsilon$, the first condition is satisfied.

In order to find F_2, we invert T and L; this requires that $\|T - L\| < 1/k$. Writing $L = T - (T - L)$, we have

$$T^{-1}L = \mathrm{id} - T^{-1}(T - L)$$

and since

$$\|T^{-1}(T - L)\| \leq k\|T - L\| < 1$$

the inverse of $T^{-1}L$ exists and is given by the power series $\sum_{j=0}^{\infty} [T^{-1}(T - L)]^j$.

Comparing $(T^{-1}L)^{-1}$ with the identity, we see

$$\|\mathrm{id} - (T^{-1}L)^{-1}\| \leq \sum_{j=1}^{\infty} [k\|T - L\|]^j = \frac{k\|T - L\|}{1 - k\|T - L\|} \leq \frac{k\delta}{1 - k\delta}.$$

Thus

$$\|T^{-1} - L^{-1}\| = \|[\mathrm{id} - (T^{-1}L)^{-1}\| \leq \frac{k^2\delta}{1 - k\delta}.$$

In order to apply the preceding arguments to L^{-1}, it suffices that

$$\frac{k^2\delta}{1 - k\delta} < \frac{1 - \tau}{2} \quad \text{and} \quad \frac{1}{1/\tau - k^2\delta/(1 - k\delta)} < \tau + \varepsilon,$$

which we may guarantee by making δ small enough. □

Proposition 7.6. *Let Λ be a hyperbolic invariant set for a C^r diffeomorphism of M, $r \geq 1$. There exist neighborhoods U of Λ in M and V of f in* $\mathrm{Diff}^1(M)$ *such that if g belongs to V and if K is a g-invariant set in U, then K is hyperbolic for g.*

Proof. Consider the hyperbolic splitting of TM over Λ:

$$TM|_\Lambda = E^s \oplus E^u, \qquad \|Df|_{E^s}\| < \lambda < 1; \qquad \|Df^{-1}|_{E^u}\| < \lambda < 1.$$

First, we continuously extend the bundles E^s and E^u over Λ to bundles $\tilde{E}^s$ and $\tilde{E}^u$ over an open neighborhood U_1; then, using the openness of linear isomorphisms, we find a neighborhood U_2 of Λ_1, contained in U_1, where $TM|_{U_2} = \tilde{E}^s \oplus \tilde{E}^u$.

If x is a point in the intersection of U_2 and $f^{-1}(U_2)$ we can write $Df\colon \tilde{E}^s_x \oplus \tilde{E}^u_x \to \tilde{E}^s_{f(x)} \oplus \tilde{E}^u_{f(x)}$ as a block matrix

$$\begin{pmatrix} A'_x & B'_x \\ C'_x & D'_x \end{pmatrix}$$

and restricting our attention to a possibly smaller neighborhood U_3 of Λ, we have

$$\|A'_x\| < \lambda + \frac{\delta}{2}; \qquad \|B'_x\| < \frac{\delta}{2}; \qquad \|C'_x\| < \frac{\delta}{2}; \qquad \|(D'_x)^{-1}\| < \lambda + \frac{\delta}{2}.$$

Now let U_4 be a neighborhood of Λ with compact closure such that $\bar{U}_4 \cup f(\bar{U}_4)$ is contained in U_3. If g is C^1 close to f, the neighborhood U_2 will contain $g(U_4)$ and, for x in U_4, the linear map $D_{g_x}\colon \tilde{E}^s_x \oplus \tilde{E}^u_x \to \tilde{E}^s_{g(x)} \oplus \tilde{E}^u_{g(x)}$ has the matrix

$$\begin{pmatrix} A_x & B_x \\ C_x & D_x \end{pmatrix}$$

with

$$\|A_x\| < \lambda + \delta < 1; \qquad \|B_x\| < \delta; \qquad \|C_x\| < \delta; \qquad \|D_x^{-1}\| < \lambda + \delta.$$

Notice that

$$\begin{pmatrix} A_x & B_x \\ C_x & D_x \end{pmatrix}$$

and

$$\begin{pmatrix} A_x & 0 \\ 0 & D_x \end{pmatrix}$$

are δ close.

If K is a g invariant set in U_4, then its unstable bundle is the fixed point of a certain graph transform. Let $D \to K$ be the bundle whose fiber at x is the unit ball in the space of linear maps from $\tilde{E}^u_x$ to $\tilde{E}^s_x$; $D_x = \mathbf{L}_1(\tilde{E}^u_x, \tilde{E}^s_x)$.

Define a fiber preserving map $F\colon D \to D$ by

$$F|_{D_x} = F_x = \Gamma_{D_{g_x}}\colon \mathbf{L}_1(\tilde{E}^u_x, \tilde{E}^s_x) \to \mathbf{L}_1(\tilde{E}^u_{g(x)}, \tilde{E}^s_{g(x)}).$$

F is continuous and contracts fibers. The graph transform Γ_F thus has a unique fixed point, as in the case of the graph transform in the proof of the C^r section theorem. We leave it to the reader to check that the invariant section for Γ_F gives the unstable bundle for g and to construct the stable bundle by considering g^{-1}. □

REMARK. One could demonstrate the existence of a unique fixed point for Γ_F differently. Let

$$L = \begin{pmatrix} A_x & 0 \\ 0 & B_x \end{pmatrix}$$

and define a fiber preserving map $G\colon D \to D$ by $G_x = \Gamma_L$. Now Γ_G is a contracting linear operator; Γ_F, being a small perturbation of Γ_G, thus has the desired fixed point. Note that L is not the derivative of a mapping, but nevertheless allows us to show that Γ_F has a fixed point. We will use this idea again to prove the stronger Theorem 7.8.

The next proposition is a consequence of the stable manifold theorem, but we will give a direct proof in order to obtain the estimates we need for the proof of Theorem 7.8.

Proposition 7.7. *Let E be a Banach space, $E(r)$ the closed ball in E about 0 of radius r. Suppose that E admits a splitting by closed subspaces $E = E_1 \oplus E_2$, and that E has the box norm. Let p_i be the projection on E_i.*

Let T be a linear automorphism which is hyperbolic with respect to the splitting $E_1 \oplus E_2$. Let $T_i = T|_{E_i}$. Thus there is a positive constant $\lambda < 1$ such that $\|T_1^{-1}\| < \lambda$ and $\|T_2\| < \lambda$.

Let $f\colon E(r) \to E$ be a map close to T with $\mathrm{Lip}(f - T) < \varepsilon$ and $\|f(0)\| \le \delta$.

If $\lambda + \varepsilon < 1$ and $\delta < r(1 - \lambda - \varepsilon)$, then f has a unique fixed point p_f in $E(r)$. Moreover,

$$\|p_f\| < \frac{1}{1 - \lambda - \varepsilon} \|f(0)\|$$

and p_f depends C^0 continuously on f.

PROOF. Define a map $\bar{f}\colon E(r) \to E$ by

$$\bar{f}(x) = T_1^{-1}[x_1 + T_1 x_1 - f_1(x_1, x_2)] + f_2(x_1, x_2),$$

where $x_i = p_i(x)$ and $f_i = p_i \circ f$. The map $\bar{f}$ has the same fixed points as f; we will show that $\bar{f}$ is contracting and $\bar{f}[E(r)] \subset E(\mathrm{r})$, so $\bar{f}$ has a unique fixed point by the contraction mapping principle. Letting $x = (x_1, x_2)$ and $y = (y_1, y_2)$ we see

$$\begin{aligned} \|\bar{f}(x) - \bar{f}(y)\| &= \max\{\|T_1^{-1}[(x_1 - y_1) + (T_1 p_1 - f_1)(x) \\ &\qquad - (T_1 p_1 - f_1)(y)]\|, \|f_2(x) - f_2(y)\|\} \\ &\le \max\{\lambda(1 + \varepsilon)\|x - y\|, (\lambda + \varepsilon)\|x - y\|\} = (\lambda + \varepsilon)\|x - y\|. \end{aligned}$$

Since $(\lambda + \varepsilon) < 1$, $\bar{f}$ is contracting.

Next we will show that $\bar{f}[E(r)]$ is contained in $E(r)$. From above, we conclude that

$$\|\bar{f}(x)\| \leq (\lambda + \varepsilon)\|x\| + \|\bar{f}(0)\|.$$

Now, since $\|T_1^{-1}\| < 1$,

$$\|\bar{f}(0)\| = \max[\|T_1^{-1}f_1(0)\|, \|f_2(0)\|] \leq \|f(0)\|.$$

Therefore, when x is in $E(r)$

$$\|\bar{f}(x)\| \leq (\lambda + \varepsilon)\|x\| + \|f(0)\| \leq (\lambda + \varepsilon)r + \delta$$

which is less than r by hypothesis. Since p_f is the limit of the sequence $\bar{f}^n(0)$, we easily see

$$\|p_f\| \leq \frac{\|\bar{f}(0)\|}{1 - (\lambda + \varepsilon)} \leq \frac{1}{1 - \lambda - \varepsilon}\|f(0)\|.$$

It only remains to show that p_f depends continuously on f, Defining $\bar{g}$ analogously to $\bar{f}$, we estimate

$$\|\bar{f}(x) - \bar{g}(x)\| \leq \|f(x) - g(x)\| \leq d(f, g).$$

Hence

$$\begin{aligned}\|p_f - p_g\| = \|\bar{f}(p_f) - \bar{g}(p_g)\| &\leq \|\bar{f}(p_f) - \bar{g}(p_g)\| + \|\bar{g}(p_f) - \bar{g}(p_g)\| \\ &\leq d(f, g) + (\lambda + \varepsilon)\|p_f - p_g\|.\end{aligned}$$

Thus we have

$$\|p_f - p_g\| \leq \frac{d(f, g)}{1 - \lambda - \varepsilon}. \qquad \square$$

Now we will examine the fundamental theorem which allows us to deduce easily most of the stability theorems in this book.

Theorem 7.8. *Let Λ be hyperbolic invariant set for a C^k diffeomorphism of M. There are numbers $\alpha > 0$, $K > 0$, $r > 0$, a neighborhood U of Λ in M and neighborhood V of f in $\mathrm{Diff}^1(M)$ with the following properties:*

For any topological space X, any homeomorphism h of X, and any continuous map $i: X \to U$, if g belongs to V and $d(ih, gi) < \alpha$, then there is a continuous map $j: X \to M$ such that $jh = gj$ and $d(i, j) \leq r$. (We call such a j a pseudoconjugacy).

In fact, we have the stronger estimate that $d(i, j) \leq Kd(ih, gi)$. Moreover, for fixed i and h, j depends C^0 continuously on g.

We can summarize this by the following diagrams:

$$\begin{array}{ccc} X & \xrightarrow{\ i\ } & U \\ {\scriptstyle h}\downarrow & & \downarrow{\scriptstyle g} \\ X & \xrightarrow[\ i\ ]{} & M \end{array}$$

commutes up to α *and* g *is close to* f. $\Rightarrow \exists! j$, $d(i, j) \le r$ *such that*

$$\begin{array}{ccc} X & \xrightarrow{j} & M \\ {\scriptstyle h}\downarrow & & \downarrow{\scriptstyle g} \\ X & \xrightarrow[j]{} & M \end{array}$$

commutes.

PROOF. Our strategy for the proof will be to show that

$$G: C^0(X, M) \to C^0(X, M),$$
$$G: k \mapsto gkh^{-1}$$

has a unique fixed point in a neighborhood of i. We will do this by showing that G is a Lipschitz perturbation of a hyperbolic operator constructed from f, and then apply Proposition 7.7.

Step 0: *Linearizing* $C^0(X, M)$. Suppose we are given a Riemannian metric on M. There is an $\varepsilon' > 0$ such that the exponential map $\exp_x$: $T_xM \to M$ is a diffeomorphism on the ball of radius ε'. We will identify the tangent space of T_xM with T_xM itself, since it is a linear space. Thus, the derivative of $\exp_x$ at y in T_xM, $(D \exp_x)(y)$ maps T_xM to $T_{\exp_x(y)}M$ and $(D \exp_x)(0)$ is the identity map of T_xM. Choose ε with $0 < \varepsilon < \varepsilon'$. The ball $\overline{B_\varepsilon(i)} = \{k: X \to M \mid d(i, k) \le \varepsilon\} \subset C^0(X, M)$ can be identified with a neighborhood $\Gamma^0_\varepsilon(X, i^*TM)$ of the zero section of the bundle i^*TM by a homeomorphism Φ defined by $\Phi(k)(x) = \exp_{i(x)}^{-1}[k(x)]$. Note that Φ sends i to the zero section.

We will assume that the given metric is adapted to (Λ, f), uniformly on Λ, that is,

$$\exists \lambda < 1$$

such that

$$\forall x \in \Lambda, \qquad \|Df_x|_{E^s}\| < \lambda$$

and

$$\|Df_x^{-1}|_{E^u}\| < \lambda.$$

Extend the splitting $T_\Lambda M = E^s \oplus E^u$ to a compact neighborhood W of Λ, $T_W M = \tilde{E}^s \oplus \tilde{E}^u$. If z and x are two points in M with $d(f(x), z) \le \varepsilon$, define a map $F_{z,x}$: $T_xM \to T_zM$ by $F_{z,x} = D(\exp_z^{-1})_{f(x)}Df_x$. If the points z and x belong to W, the splitting $\tilde{E}^s \oplus \tilde{E}^u$ allows us to write $F_{z,x}$ as the block matrix

$$\begin{pmatrix} A_{z,x} & B_{z,x} \\ C_{z,x} & D_{z,x} \end{pmatrix}$$

where

$$A_{z,x} \in \mathbf{L}(\tilde{E}^s_x, \tilde{E}^s_z), \qquad B_{z,x} \in \mathbf{L}(\tilde{E}^u_x, \tilde{E}^s_z),$$
$$C_{z,x} \in \mathbf{L}(\tilde{E}^s_x, \tilde{E}^u_z), \quad \text{and} \quad D_{z,x} \in \mathbf{L}(\tilde{E}^u_x, \tilde{E}^u_z).$$

Let $\tilde{F}_{z,x}$ be the map with the diagonal block matrix

$$\begin{pmatrix} A_{z,x} & 0 \\ 0 & D_{z,x} \end{pmatrix}.$$

In this setting we prove two preliminary lemmas before attacking the proof of Theorem 7.8. □

Lemma 7.9. *For all $\eta > 0$ we find a neighborhood $U(\eta)$ of Λ and a constant $\delta > 0$ such that*

$$\forall x \in U(\eta), \qquad \forall z \quad \textit{such that} \quad d(f(x), z) < \delta,$$

$$\|F_{z,x} - \tilde{F}_{z,x}\| < \eta, \qquad \|\tilde{F}_{z,x}|_{\tilde{E}^s_x}\| < \lambda, \quad \text{and} \quad \|\tilde{F}^{-1}_{z,x}|_{\tilde{E}^u_x}\| < \lambda.$$

Proof. If x belongs to Λ and $z = f(x)$ we have $F_{z,x} = \tilde{F}_{z,x} = Df_x$. For arbitrary x and z the result follows from the continuous dependence of $F_{z,x}$ and $\tilde{F}_{z,x}$ on z and x. □

Lemma 7.10. *For all η' and $\delta, \eta' > 0$, $0 < \delta < \varepsilon$. there is $r = r(\delta, \eta') > 0$ and a C^1 neighborhood $V = V(\delta, \eta')$ of f such that if $d(z, f(y)) \leq \delta$ and g belongs to V, then we have*

$$g(B_r(y)) \subset \exp_z(B_\varepsilon(0)), \tag{1}$$

$$\operatorname{Lip}[(F_{z,y} - \exp_z^{-1} \circ g \circ \exp_y)|_{B_r(0)}] < \eta'. \tag{2}$$

Proof. First choose a small $r_1 = r_1(\delta)$ so that if $d(z, f(y)) < \delta$ we have $\exp_z(B_\varepsilon(0)) \supset f(\bar{B}_{r_1}(y))$, which is possible since $\delta < \varepsilon$ and f is uniformly continuous. Condition (1) then holds for g sufficiently C^0 close to f.

The map $\exp_z^{-1} f \exp_y$ is defined on the ball $B_{r_1}(0)$; if $\exp_z^{-1}$ is also defined in a neighborhood of a point v within this ball, we have

$$\begin{aligned} D(F_{z,y} &- \exp_z^{-1} g \exp_y)_v \\ &= D[(D\exp_z^{-1})_{f(y)} Df_y - \exp_z^{-1} g \exp_y]_v \\ &= (D\exp_z^{-1})_{f(y)} Df_y - D(\exp_z^{-1})_{g(\exp_y(v))} Dg_{\exp_y(v)} D(\exp_y)_v. \end{aligned}$$

This last expression is defined and equal to 0 if $g = f$ and $v = 0$. Since the derivative above is a continuous function of v in $T_x M$ and g in $\operatorname{Diff}^1(M)$, we are done, by continuity. □

We now continue with our proof of the theorem.

Step 1. We will construct a hyperbolic linear operator $\mathbf{F}$ on $\Gamma(X, i^*TM)$ which depends only on f, i, and h. We define $\mathbf{F}$ by the formula

$$\mathbf{F}(\sigma)(x) = \tilde{F}_{i(x), ih^{-1}(x)} \sigma(h^{-1}(x)), \qquad \text{where} \quad \sigma \in \Gamma(X, i^*TM).$$

Since $\tilde{F}_{z,x}$ is only defined where x and z are in a small neighborhood of Λ and the distance from $f(x)$ to z is small, the operator $\mathbf{F}$ is well defined only when

$i(x)$ is contained in a small neighborhood of Λ and $d(i, fih^{-1})$ is small. Notice that, by construction, $\mathbf{F}$ preserves the splitting $\Gamma(X, i^*TM) = \Gamma(X, i^*\tilde{E}^u) \oplus \Gamma(X, i^*\tilde{E}^s)$.

In particular, assuming the neighborhood of Λ and the distance $d(i, fih^{-1})$ are perhaps even smaller, Lemma 7.9 shows that $\mathbf{F}$ is hyperbolic, that is,

$$\|\mathbf{F}|_{\Gamma(X, i^*\tilde{E}^s)}\| < \lambda < 1,$$

$$\|\mathbf{F}^{-1}|_{\Gamma(X, i^*\tilde{E}^u)}\| < \lambda < 1.$$

Lemma 7.9 also allows us to associate to each $\eta > 0$ a neighborhood $U(\eta)$ of Λ and a number $\delta(\eta)$ such that

$$d(i, fih^{-1}) < \delta(\eta) \quad \text{and} \quad i(X) \subset U(\eta)$$

imply

$$\sup_{x \in X} \|F_{i(x), ih^{-1}(x)} - \tilde{F}_{i(x), ih^{-1}(x)}\| < \eta.$$

Step 2: Defining G and $\mathbf{G}$. Consider the map $G: C^0(X, M) \to C^0(X, M)$, $G: k \mapsto gkh^{-1}$. From the triangle inequality

$$d(gkh^{-1}, i) \le d(gkh^{-1}, gih^{-1}) + d(gih^{-1}, i)$$

we see that there are constants $\alpha_1 > 0$, $r_1 > 0$, and a C^0 neighborhood V_1 of f (independent of X, i and k) such that if g belongs to V_1 and $d(gi, ih) < \alpha_1$, then the image $G(B_{r_1}(i))$ is contained in $B_\varepsilon(i)$.

Thus, when these two conditions hold we can define the map $\mathbf{G} = \Phi G \Phi^{-1}: \Gamma_{r_1}(X, i^*TM) \to \Gamma(X, i^*TM)$. Recall that Φ is the following chart on a neighborhood of i in $C^0(X, M)$

$$\Phi: B_\varepsilon(i) \to \Gamma_\varepsilon(X, i^*TM),$$

$$\Phi(k)(x) = \exp_{i(x)}^{-1} k(x).$$

$\mathbf{G}$ is given explicitly as

$$\mathbf{G}(\sigma)(x) = [\exp_{i(x)}^{-1}]g(\exp_{i(h^{-1}(x))}\sigma(h^{-1}(x))).$$

Step 3: $\mathbf{G}$ *is Lipschitz close to* $\mathbf{F}$. Using the norm on $\Gamma(X, i^*TM)$ induced by the Riemannian metric of M, we can calculate the Lipschitz distance from $\mathbf{G}$ to $\mathbf{F}$ on the ball $B(r')$:

$$\begin{aligned}\mathrm{Lip}[(\mathbf{G} - \mathbf{F})|_{B(r')}] \le {} & \sup_{x \in X} \|\tilde{F}_{i(x), ih^{-1}(x)} - F_{i(x), ih^{-1}(x)}\| \\ & + \sup_{x \in X} \mathrm{Lip}[F_{i(x), ih^{-1}(x)} - \exp_{ix}^{-1} g \exp_{ih^{-1}(x)}|_{B_{r'}(O_{ih^{-1}(x)})}].\end{aligned}$$

Now, from Step 1, given η we can find a constant $\delta(\eta)$ and a neighborhood $U(\eta)$ such that if $i(x)$ is contained in $U(\eta)$ and $d(i, fih^{-1}) < \delta(\eta)$, then

$$\sup_{x \in X} \|\tilde{F}_{i(x), ih^{-1}(x)} - F_{i(x), ih^{-1}(x)}\| < \eta.$$

We can also find a constant $\alpha_2(\eta)$ and a C^0 neighborhood $V_2(\eta)$ of f such that if $d(gi, ih) < \alpha_2(\eta)$ and g belongs to $V_2(\eta)$, then $d(i, fih^{-1}) < \delta(\eta)$.

Now, given $\eta' > 0$, Lemma 7.10 allows us to find a constant $r(\eta', \delta(\eta))$ and a neighborhood $V_3(\eta', \delta(\eta))$ of f such that if g belongs to $V_3(\eta', \delta(\eta))$ and $d(i, fih^{-1}) < \delta(\eta)$) (which will be the case if g also belongs to $V_2(\eta)$ and we have $d(gi, ih) < \alpha_2(\eta)$ and $r' \le r(\eta', \delta(\eta))$), then

$$\sup_{x\in X} \mathrm{Lip}[F_{i(x), ih^{-1}(x)} - \exp_{i(x)}^{-1} g \exp_{ih^{-1}(x)}|_{B_{r'}(O_{ih^{-1}(x)})}] < \eta'.$$

Summarizing, if α, V, and U satisfy

$$\alpha < \inf(\alpha_1, \alpha_2(\eta)),$$
$$r' \le r(\eta', \delta(\eta)),$$
$$V \subset V_1 \cap V_2(\eta) \cap V_3(\eta', \delta(\eta)),$$
$$U \subset U(\eta),$$

then, whenever $d(gi, ih) < \alpha$, g belongs to V, and $i(x)$ is contained in U, we have

$$\mathrm{Lip}[(\mathbf{G} - \mathbf{F})|_{B(r')}] \le \eta + \eta',$$
$$\|\mathbf{G}(0)\| = d(gih^{-1}, i) < \alpha,$$
$$\|\mathbf{F}|_{\Gamma(x, i^*\tilde{E}^s)}\| < \lambda < 1 \quad \text{and} \quad \|\mathbf{F}^{-1}|_{\Gamma(x, i^*\tilde{E}^u)}\| < \lambda < 1.$$

Step 4: The fixed point of **G**. We wish to apply Proposition 7.7. In order to do so we must use the box norm on $\Gamma(x, i^*TM) = \Gamma(x, i^*\tilde{E}^s) \oplus \Gamma(x, i^*\tilde{E}^u)$. There is a constant c, which only depends on the extensions of E^s and E^u to the neighborhood W of Λ, which allows us to see the equivalence of $\|\ \|_{\text{box}}$ and $\|\ \|_{\text{riem}}$ on $E^u \oplus E^s$:

$$1/c\|\ \|_{\text{box}} \le \|\ \|_{\text{riem}} \le c\|\ \|_{\text{box}};$$

and similarly for the associated norms on $\Gamma(X, i^*TM) = \Gamma(X, i^*\tilde{E}^u) \oplus \Gamma(X, i^*\tilde{E}^s)$.

We can thus rewrite the estimates of Step 3:

$$\mathrm{Lip}_{\text{box}}[(\mathbf{G} - \mathbf{F})|_{B(r')}] \le c^2(\eta + \eta') \quad \text{and} \quad \|\mathbf{G}(0)\|_{\text{box}} < c\alpha.$$

If r'' is less than r'/c, the box $E_1(r'') \times E_2(r'')$ is contained in the Riemannian ball $B(r')$, and we have

$$\mathrm{Lip}_{\text{box}}[(\mathbf{G} - \mathbf{F})|_{E_1(r'')\times E_2(r'')}] \le \mathrm{Lip}_{\text{box}}[(\mathbf{G} - \mathbf{F})|_{B(r')}] \le c^2(\eta + \eta').$$

In order to apply Proposition 7.7, we must have

$$(*) \qquad \lambda + c^2(\eta + \eta') < 1$$

and

$$(**) \qquad c\alpha < r''[1 - \lambda - c^2(\eta + \eta')].$$

Therefore, we first choose η and η' satisfying $(*)$. Then, we find the neighborhoods U of Λ and V of f, constants r' and $r'' < r'/c$, as above; finally we choose $\alpha < \operatorname{Inf}(\alpha_1, \alpha_2(\eta))$ small enough so that $(**)$ holds.

Proposition 7.7 then gives the existence of a pseudoconjugacy j of g and h $(gj = jh)$ satisfying $\|\Phi(j)\|_{\text{box}} \leq r''$. More precisely, we have the estimate

$$\|\Phi(j)\|_{\text{box}} \leq [1 - \lambda - c^2(\eta + \eta')]^{-1} \|\mathbf{G}(0)\|_{\text{box}}.$$

Taking $r < r''/c$, we see the inequality $d(i, k) < r$ implies $\|\Phi(k)\|_{\text{box}} \leq r''$. Thus, if j is a solution of the pseudoconjugacy equation, satisfying $d(i, j) \leq r$, then j is the only such solution. On the other hand, since

$$d(i, j) \leq [c^2/(1 - \lambda - c^2(\eta + \eta'))] d(gi, ih),$$

when α is small enough, the solution j does, in fact, satisfy $d(i, j) \leq r$.

Furthermore, if i and h are fixed, j depends continuously on $\mathbf{G}$, which, in turn, depends continuously on g, since

$$\|\mathbf{G}_1 - \mathbf{G}_2\|_{\text{box}} \leq c \|\mathbf{G}_1 - \mathbf{G}_2\|_{\text{riem}} \leq \text{cst.}\ d(G_1, G_2) \leq \text{cst.}\ d(g_1, g_2). \qquad \square$$

FINAL REMARKS. In case the reader has difficulty seeing the forest for the trees above, we sketch a more comprehensible heuristic proof.

Consider the map

$$\mathbf{G}\colon \Gamma_r^0(X, i^*TM) \to \Gamma^0(X, i^*TM),$$

where

$$\mathbf{G}(\sigma)(x) = (\exp_{i(x)}^{-1}) g(\exp_{ih^{-1}(x)} \sigma h^{-1}(x)).$$

This has as its derivative at the zero section

$$\mathbf{G}'\colon \Gamma^0(X, i^*TM) \to \Gamma^0(X, i^*TM),$$

where

$$\mathbf{G}'(\sigma)(x) = \pi D(\exp_{i(x)}^{-1})_{gih^{-1}(x)} D_{g_{ih^{-1}(x)}} \sigma h^{-1}(x).$$

When g is close to f, we see, as in Lemma 7.10, that $\mathbf{G}'$ is close to $\mathbf{F}$ and thus $\mathbf{G}'$ is hyperbolic. If $\mathbf{G}(0)$ is near 0, then, in a neighborhood of 0, $\mathbf{G}$ is a Lipschitz perturbation of its derivative at 0. In this case, $\mathbf{G}$ has a fixed point near 0. However, the norm of $\mathbf{G}(0)$ is bounded as follows:

$$\|\mathbf{G}(0)\| = d(\mathbf{G}(0), i) = d(gih^{-1}, i) = d(gi, ih).$$

We conclude the chapter with the linearization theorem of Hartman and Grobman, which says that, near a hyperbolic fixed point, a diffeomorphism is conjugate to its linear part.

When E is a Banach space, we denote by $C_b^0(E, E)$ the space of bounded, continuous functions from E to itself. $C_b^0(E, E)$ is a Banach space with the sup norm.

Theorem 7.11. *Let $L: E \to E$ by a hyperbolic operator; a small Lipschitz perturbation of L is topologically conjugate to L. More precisely, given a hyperbolic operator L, there is an $\varepsilon > 0$ such that for any $\Phi \in C_b^0(E, E)$ of Lipschitz norm less than or equal to ε, there is a homeomorphism H conjugating $L + \Phi$ and L: $L + \Phi = HLH^{-1}$.*

PROOF. Suppose that E is split as $E = E^u \oplus E^s$ and E has the box norm relative to this splitting. Since L is hyperbolic, there is a λ, $0 < \lambda < 1$, such that

$$\|L^{-1}|_{E^u}\| < \lambda \quad \text{and} \quad \|L|_{E^s}\| < \lambda.$$

The homeomorphism H, if it exists must satisfy $(L + \Phi)H = HL$.

We will prove an apparently stronger result: there is an $\varepsilon > 0$ such that if two maps $\Phi, \Phi' \in C_b^0(E, E)$ are Lipschitz and ε small in the Lipschitz metric ($\text{Lip}\,\Phi < \varepsilon$ and $\text{Lip}\,\Phi' < \varepsilon$), then there is a unique map g in $C_b^0(E, E)$ for which

$$(***) \qquad (L + \Phi)(\text{id} + g) = (\text{id} + g)(L + \Phi').$$

Exchanging Φ and Φ' and using the uniqueness of g we can easily see that $\text{id} + g$ is a homeomorphism.

Note that if $\varepsilon \leq \|L^{-1}\|^{-1}$, the implicit function theorem guarantees that $L + \Phi$ and $L + \Phi'$ are homeomorphisms.

The equation (***) is equivalent, then, to

$$(L + \Phi)(\text{id} + g)(L + \Phi')^{-1} = \text{id} + g,$$

or, expanding

$$Lg(L + \Phi')^{-1} + \Phi g(L + \Phi')^{-1} + (L + \Phi)(L + \Phi')^{-1} - \text{id} = g.$$

We define a map

$$L_{\Phi'}^*: C_b^0(E, E) \to C_b^0(E, E)$$

by

$$L_{\Phi'}^*(g) = Lg(L + \Phi')^{-1}$$

and another map

$$\Phi^*: C_b^0(E, E) \to C_b^0(E, E)$$

by

$$\Phi^*(g) = \Phi g(L + \Phi')^{-1}.$$

Setting $k = (L + \Phi)(L + \Phi')^{-1} - \text{id}$, we see that to find g we must find a fixed point of the map $g \mapsto L_{\Phi'}^* g + \Phi^* g + k$. We will find the fixed point by applying Proposition 7.7.

First we must check the hypotheses:

First, the function k belongs to $C_b^0(E, E)$. In fact, setting $(L + \Phi')^{-1} = L^{-1} + w$, we have $(L^{-1} + w)(L + \Phi') = \text{id}$, so $w = -L^{-1}\Phi'(L + \Phi')^{-1}$ and w

belongs to $C_b^0(E, E)$; however, we also have

$$k = (L + \Phi)(L^{-1} + w) - \mathrm{id} = Lw + \Phi(L^{-1} + w) \in C_b^0(E, E).$$

Thus the map $g \mapsto L_{\Phi'}^* g + \Phi^* g + k$ sends $C_b^0(E, E)$ into itself.

Next, the map $L_{\Phi'}^*$ is hyperbolic. The restriction of $L_{\Phi'}^*$ to $C_b^0(E, E^u)$ is an expansion and the restriction to $C_b^0(E, E^s)$ is a contraction. The reader can easily check that

$$\|L_{\Phi'}^{*-1}|_{C_b^0(E,E^u)}\| < \lambda < 1 \quad \text{and} \quad \|L_{\Phi'}^*|_{C_b^0(E,E^s)}\| < \lambda < 1.$$

Thirdly, the map Φ^* is Lipschitz and satisfies $\mathrm{Lip}\, \Phi^* < \mathrm{Lip}\, \Phi$. We have, in fact, the estimate

$$\begin{aligned}\|\Phi^* g - \Phi^* h\| &= \sup_{x \in E} \|\Phi g(L + \Phi')^{-1} x - \Phi h(L + \Phi')^{-1} x\| \\ &\le \mathrm{Lip}\, \Phi \sup_{x \in E} \|g(L + \Phi')^{-1} x - h(L + \Phi')^{-1} x\| \\ &= \mathrm{Lip}\, \Phi \|g - h\|.\end{aligned}$$

The remaining hypotheses of Proposition 7.7 are automatically satisfied, since $r = +\infty$ here. Therefore, there is an $\varepsilon > 0$, depending only on λ, such that if $\mathrm{Lip}\, \Phi < \varepsilon$, $L_\Phi^* + \Phi^* + k$ has a unique fixed point.

We have shown, then, that if Φ and Φ' are Lipschitz small enough, then there is a unique map g in $C_b^0(E, E)$ such that

$$(L + \Phi)(\mathrm{id} + g) = (\mathrm{id} + g)(L + \Phi'). \qquad \square$$

Note that a Lipschitz map $\Phi\colon E(r) \to E$ can always be extended to a Lipschitz map $\tilde{\Phi}\colon E \to E$ with $\mathrm{Lip}\, \tilde{\Phi} \le 2\, \mathrm{Lip}\, \Phi$ as follows:

$$\tilde{\Phi}(x) = \Phi(x) \quad \text{if} \quad \|x\| \le r,$$

$$\tilde{\Phi}(x) = \Phi\left(r \frac{x}{\|x\|}\right) \quad \text{if} \quad \|x\| \ge r.$$

From this we can deduce the following corollary.

Corollary 7.12. *Let L be a hyperbolic automorphism of E. There is an $\varepsilon > 0$ such that if $f\colon E(r) \to E$ satisfies $\mathrm{Lip}(f - L) < \varepsilon$ and $f(0) = 0$, then there is a local homeomorphism h on a neighborhood of 0 such that $h(0) = 0$ and $hf = Lh$ near 0.* □

Since a smooth map is a small Lipschitz perturbation of its derivative at a fixed point we also have:

Theorem 7.13 (Grobman-Hartman). *If p is a hyperbolic fixed point of a C^1 diffeomorphism f of M, then f is topologically conjugate to its derivative in a neighborhood of p.* □

EXERCISE 7.1. Let Λ be a closed, hyperbolic invariant set for $f \in \mathrm{Diff}^k(M)$, $k \geq 1$. Show directly from Theorem 7.8 that f is expansive on Λ. More precisely, show that the constant r given by the theorem is a constant of expansivity for Λ.

EXERCISE 7.2. If p is a hyperbolic periodic point of $f \in \mathrm{Diff}^1(M)$, and $g \in \mathrm{Diff}^1(M)$ is sufficiently C^1 close to f, then g has a periodic point near p. (*Hint*: apply Proposition 7.7). From this deduce a quick proof of Theorem 7.8, in the case where Λ is finite.

EXERCISE 7.3. Prove in detail, that the solenoid, as in Chapter 4, is a closed hyperbolic set.

EXERCISE 7.4. If p is a fixed point for $f: M \to M$ and $Df(p)$ has a three-way splitting at p, $E^s \oplus E^c \oplus E^u$ as in Theorem III.7, prove that there are neighborhoods U of p in M and V of p in $E^s \times W^c_{\mathrm{loc}}(p) \times E^u$, and a homeomorphism $h: V \to U$ such that

$$fh = h(Df|E^s \times f|W^c_{\mathrm{loc}}(p) \times Df|E^u).$$

That is, the hyperbolic part of f can always be linearized. The proof of this proposition is not as easy as the usual exercise. Understanding the geometric version of the Grobman Hartman theorem in the commentaries would be helpful.

Commentary

Expansivity played a role in Smale's proof of the Ω-Stability Theorem. He asked me if I could establish it, and I made the calculation for him (see [1.16]). The stable manifold theory was not well understood in those days. The proof I give here is that of Bowen [1.2].

Definition 7.3 comes from [4.2]. Theorem 7.8 is due to Anosov. Ralph Abraham showed Anosov Mather's proof of Anosov's stability theorem during the summer of 1967 and Anosov gave a new proof close to that of Mather. Abraham sent a copy of this to Smale, who showed it to me. We will reap its benefits in the following chapters. Proposition 7.7 is a particular case of the implicit function theorem. The proof we give furnishes the estimates we need for Theorem 7.8.

The version we give of the Grobman–Hartman Theorem [7.1], our Theorem 7.13, is from [7.3] and [7.4]. I will now sketch a geometric proof of the Grobman–Hartman Theorem in finite dimensions inspired by [2.3]. We will extend the stable and unstable manifolds to a neighborhood of the hyperbolic set (cf. the commentaries on Chapter 5).

Suppose, first, that we are given two discs D_1 and D_2 of dimension k and two diffeomorphisms $f: D_1 \to \mathrm{int}\, D_1$ and $g: D_2 \to \mathrm{int}\, D_2$ which are contractions and have the same orientation (Figure C7.1). The shaded regions are annuli, that is they are diffeomorphic to $S^{k-1} \times I$ (this is a standard result in differential geometry [7.2]).

If we wish to construct a homeomorphism $h: D_1 \to D_2$ satisfying $hf = gh$, we can begin as follows. Suppose first that the restriction $h: \partial D_1 \to \partial D_2$ is any well-defined orientation preserving diffeomorphism. The restriction of h to

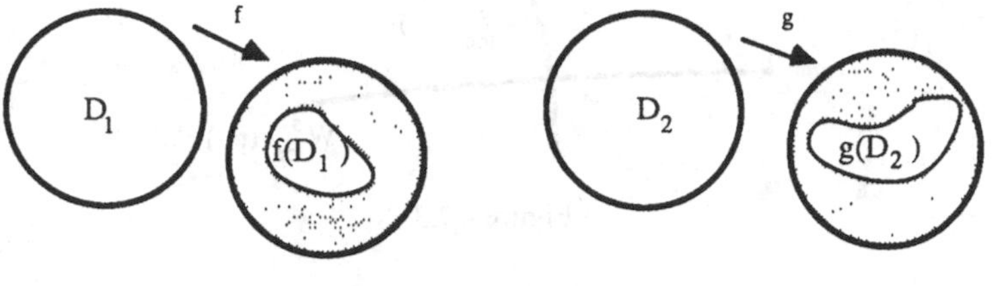

Figure C7.1.

$f(\partial D_1)$ is then defined, since we must have $h(y) = ghf^{-1}(y)$ if h is to conjugate f and g.

Now there is no obstruction to extending h to a diffeomorphism which sends the annulus A_1 bounded by ∂D_1 and $f(\partial D_1)$ onto the annulus B_1 bounded by ∂D_2 and $g(\partial D_2)$. In fact, the only obstruction can come from the orientation of the restrictions of f and g to ∂D_1 and ∂D_2.

Set $f^{i-1}(A_1) = A_i$ and $g^{i-1}(B_1) = B_i$, as shown in Figure C7.2. Since the diffeomorphism $h\colon A_1 \to B_1$ has been defined on the first annulus A_1, it is also defined on subsequent annuli by $h(y) = g^i hf^{-i}(y)$, since h must conjugate f and g. Finally, if ρ is the unique fixed point of f and q of g, then we set $h(\rho) = q$. It is not very hard now to check that h is a homeomorphism.

Now suppose that we are given two hyperbolic fixed points ρ_1 for f_1 and ρ_2 for f_2 satisfying $\dim W^s(\rho_1) = \dim W^s(\rho_2)$ and $\dim W^u(\rho_1) = \dim W^u(\rho_2)$. Suppose further that the orientations of the restriction of f_1 to $W^s(\rho_1)$ and $W^u(\rho_1)$, and of f_2 to $W^s(\rho_2)$ and $W^u(\rho_2)$ coincide. Then there will be a homeomorphism h defined on a neighborhood of ρ_1 satisfying $hf_1 = f_2 h$. The dimensions of the stable and unstable manifolds for f and the orientations of the restrictions of f to each of these manifolds thus forms a complete system of invariants for a hyperbolic fixed point.

Here is the idea of the proof one will find in [2.3]. The figures depict the case when the dimension of M is two. We are going to try to represent f_1 in a neighborhood of ρ_1 as a "product" $f_1|_{W^s_{\text{loc}}(\rho_1)} \times f_1|_{W^u_{\text{loc}}(\rho_1)}$ and f_2 in a neighborhood of ρ_2 as $f_2|_{W^s_{\text{loc}}(\rho_2)} \times f_2|_{W^u_{\text{loc}}(\rho_2)}$.

The reasoning that we have given shows that the restriction $f_1|_{W^s_{\text{loc}}(\rho_1)}$ and $f_2|_{W^s_{\text{loc}}(\rho_2)}$ are conjugate, as are the restrictions $f_1|_{W^u_{\text{loc}}(\rho_1)}$ and $f_2|_{W^u_{\text{loc}}(\rho_2)}$, since this

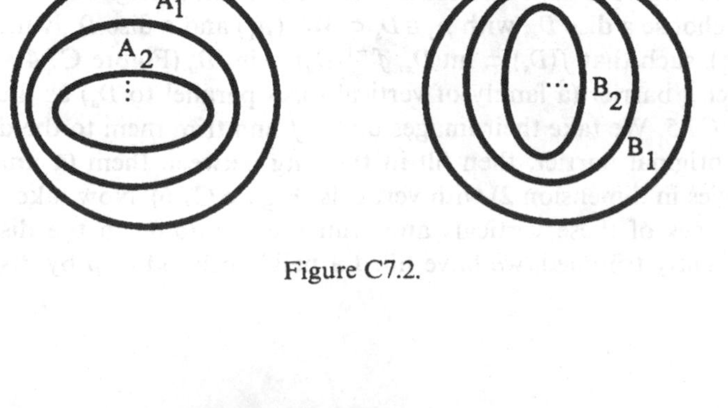

Figure C7.2.

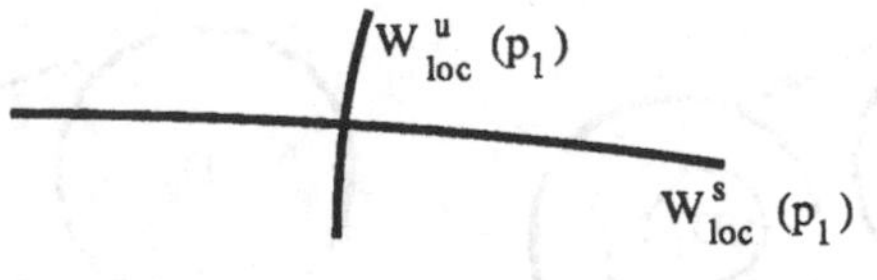

Figure C7.3.

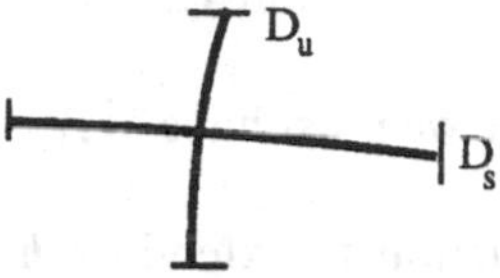

Figure C7.4.

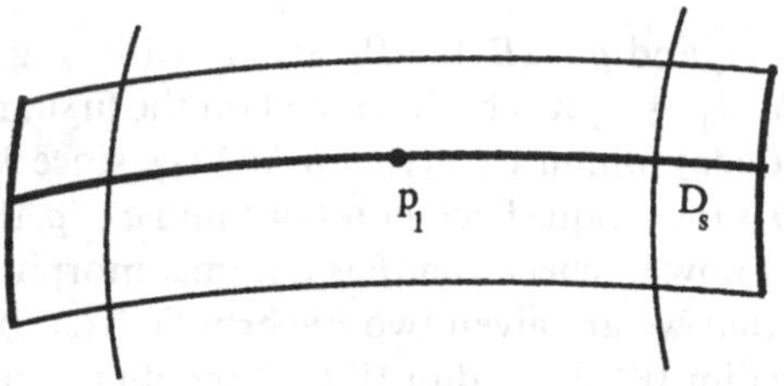

Figure C7.5.

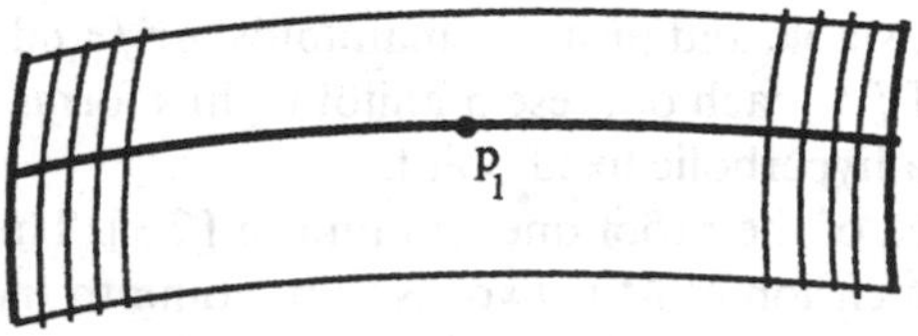

Figure C7.6.

is certainly the case for their inverses. The "product" of these conjugacies will be the sought for conjugacy of f_1 and f_2.

$W^u_{loc}(\rho_1)$ and $W^s_{loc}(\rho_1)$ intersect transversally in ρ_1. The local figure is as shown in Figure C7.3. The situation is the same in a neighborhood of ρ_2.

We choose a disc D_s, with $\rho_1 \in D_s \subset W^s_{loc}(\rho_1)$ and a disc D_u, with $\rho_1 \in D_u \subset W^u_{loc}(\rho_1)$, such that $f(D_s) \subset \text{int } D_s$, $f^{-1}(D_u) \subset \text{int } D_u$ (Figure C7.4). Along ∂D_s we erect a barrier (a family of vertical discs parallel to D_u) as illustrated in Figure C7.5. We take their images under f and trim them to the dimensions of the original barrier, then fill in the ring between them (a union of two rectangles in dimension 2) with verticals (Figure C7.6). Now take the successive images of these verticals and trim them. Adding in the disc D_u, also conveniently trimmed, we have filled a neighborhood of ρ by disjoint discs

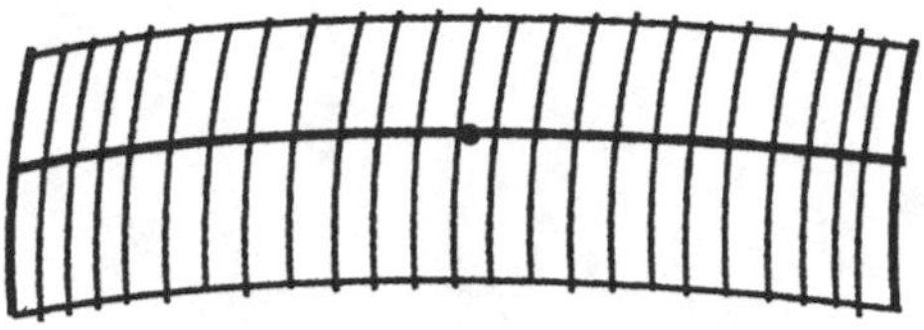

Figure C7.7.

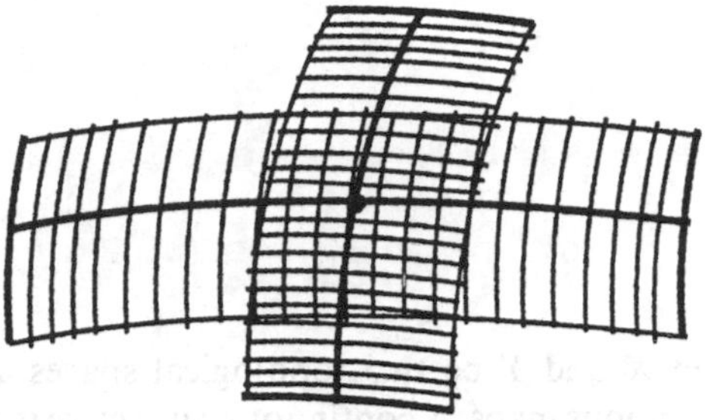

Figure C7.8.

parametrized by D_s. This defines the "x coordinate" (Figure C7.7). We also define the "y coordinate" similarly, using f_1^{-1} (Figure C7.8). This process is the geometric content of numerous linearization and stability arguments.

The method for finding the fixed point in Proposition 7.7 comes from [4.2]. Exercise 7.4 may be derived from [7.5] and in fact it comes from there.

References

[7.1] Hartman, P., *Ordinary Differential Equations*, 3rd edn, Birkhauser, Boston, 1983, 250.

[7.2] Palais, R., Extending diffeomorphisms, *Proc. Amer. Math. Soc.* **11** (1960), 274.

[7.3] Palis, J., On the local structure of hyperbolic points in Banach space, *An. Acad. Brasil. Ciênc.* **40** (1968), 263.

[7.4] Pugh, C., On a theorem of P. Hartman, *Amer. J. Math.* **91** (1969), 363

[7.5] Pugh, C. and Shub, M., Linearization of normally hyperbolic diffeomorphisms and flows, *Invent. Math.* **10** (1970), 187.

CHAPTER 8

Stability

Definition 8.1. Let X and Y be two topological spaces and $f: X \to X$ and $g: Y \to Y$ two continuous maps. A continuous, surjective map $h: X \to Y$ is said to be a *semiconjugacy* if $hf = gh$, that is if the following diagram commutes:

$$\begin{array}{ccc} X & \xrightarrow{f} & X \\ {\scriptstyle h}\downarrow & & \downarrow{\scriptstyle h} \\ Y & \xrightarrow{g} & Y. \end{array}$$

If, moreover, h is a homeomorphism of X onto Y, then h is called a *topological conjugacy*, and f and g are said to be *topologically conjugate.*

Note that $hf = gh$ implies $hf^n = g^n h$. In fact, $hf^2 = (hf)f = g(hf) = g^2 h$, etc. The image of an f orbit by a semiconjugacy is thus a g orbit, while a topological conjugacy sends f orbits to g orbits and preserves their topological properties.

The conjugacy h may be interpreted as a continuous change of variables which identifies f and g. One easily sees that topological conjugacy is an equivalence relation.

Definition 8.2. Let E be an equivalence relation on $\mathrm{Diff}^r(M)$. A map g in $\mathrm{Diff}^r(M)$ is said to be *E-stable* if the equivalence class of g contains g in its interior. In this case there is a neighborhood U of g in $\mathrm{Diff}^r(M)$ such that all the maps h in U are E-equivalent to g.

The notion of stability associated to topological conjugacy is called *structural stability.*

We can define four notions of stability corresponding to the four invariant sets which we defind in Chapter 1:

$$\overline{\mathrm{Per}(f)} \subset L(f) \subset \Omega(f) \subset R(f).$$

We say that f and g are $\overline{\mathrm{Per}}$-equivalent if the restrictions of f and g to $\overline{\mathrm{Per}(f)}$ and $\overline{\mathrm{Per}(g)}$ are topologically conjugate; and similarly for L-, Ω-, and R-equivalence.

A map stable with respect to $\overline{\mathrm{Per}}$-, L-, Ω-, or R-equivalence is said to be $\overline{\mathrm{Per}}$-, L-, Ω-, or R-stable. We have, corresponding, to the inclusions above, the implications

$$\textit{structural stability} \Rightarrow \textit{R-stability} \Rightarrow \Omega\textit{-stability}$$
$$\Rightarrow \textit{L-stability} \Rightarrow \overline{\mathrm{Per}}\textit{-stability}.$$

It is known that R-stability does not imply structural stability in general.

Let f be a C^r diffeomorphism of M. Let $N_n(f)$ be the number of fixed points of f^n. Two homeomorphisms f and g are said to be ζ-equivalent if, for all strictly positive n, $N_n(f) = N_n(g)$. Clearly $\overline{\mathrm{Per}}$-stability, implies ζ-stability; it is not known however if the converse holds, nor if ζ-stability implies L-, Ω-, or R-stability.

We will now study the case when M (resp. $R(f)$, $\Omega(f)$, $L(f)$, or $\overline{\mathrm{Per}(f)}$) is a hyperbolic invariant set for f. We begin by proving the following very general theorem:

Theorem 8.3. *Let Λ be a closed hyperbolic invariant set for a C^k diffeomorphism of M, $k \geq 1$. There is a neighborhood U_f of f in* $\mathrm{Diff}^k(M)$ *and a continuous function Φ: $U \to C^0(\Lambda, M)$ such that*:

(1) *$\Phi(f)$ is in the inclusion,* inc_Λ, *of Λ in M.*
(2) *$\Phi(g)(\Lambda)$ is a hyperbolic invariant set for g, for any g in U_f.*
(3) *$\Phi(g)$ is a homeomorphism of Λ onto $\Phi(g)(\Lambda)$ and topologically conjugates the restriction of f to Λ to the restriction of g to $\Phi(g)(\Lambda)$. In other words, the following diagram of homeomorphisms commutes*:

$$\begin{array}{ccc} \Lambda & \xrightarrow{\Phi(g)} & \Phi(g)(\Lambda) \\ {\scriptstyle f}\downarrow & & \downarrow{\scriptstyle g} \\ \Lambda & \xrightarrow{\Phi(g)} & \Phi(g)(\Lambda) \end{array}$$

(4) *There is a constant K such that $d_{C^0}(\Phi(g), \mathrm{inc}_\Lambda) < K d_{C^0}(g, f)$.*

Proof. Take neighborhoods U of Λ and V of f and constants α, r, and K as in Theorem 7.8. If g satisfies $d(f, g) < \alpha$, the following diagram commutes to α:

$$\begin{array}{ccc} \Lambda & \xrightarrow{\mathrm{inc}_\Lambda} & U \\ {\scriptstyle f}\downarrow & & \downarrow{\scriptstyle g} \\ \Lambda & \xrightarrow{\mathrm{inc}_\Lambda} & M. \end{array}$$

Thus the two conditions $g \in V$, $d(f, g) < \alpha$ guarantee the existence of a unique map $\Phi(g)\colon \Lambda \to M$ satisfying $d(\text{inc}_\Lambda, \Phi(g)) \le r$ and $g\Phi(g) = \Phi(g)f$.

Furthermore, we know that $\Phi(g)$ depends continuously on g and satisfies $d(\text{inc}_\Lambda, \Phi(g)) \le Kd(f, g)$.

Since $\Phi(f)$ and inc_Λ both make the following diagram commute:

$$\begin{array}{ccc} \Lambda & \longrightarrow & U \\ {\scriptstyle f}\big\downarrow & & \big\downarrow{\scriptstyle f} \\ \Lambda & \longrightarrow & M, \end{array}$$

we see that $\Phi(f) = \text{inc}_\Lambda$.

Theorem 7.6 implies that for g sufficiently close to f, the set $\Phi(g)(\Lambda)$ is hyperbolic.

It remains to show that $\Phi(g)$ is injective. It is possible to do this using the expansivity of f on Λ, but we will use another method which appeals to Theorem 7.8.

Suppose that g is sufficiently close to f that the set $\Lambda_g = \Phi(g)(\Lambda)$ is contained in U. Since

$$\begin{array}{ccc} \Lambda_g & \xrightarrow{\text{id}} & U \\ {\scriptstyle g}\big\downarrow & & \big\downarrow{\scriptstyle f} \\ \Lambda_g & \xrightarrow{\text{id}} & M \end{array}$$

commutes up to $d(f, g)$, there is a map j such that $jg = fj$ and $d(\text{id}|_{\Lambda_g}, j) \le Kd(f, g)$. We also see that

$$d(\text{inc}_\Lambda, j\Phi(g)) \le 2Kd(f, g),$$
$$[j\Phi(g)] \circ f = f \circ [j\Phi(g)].$$

If g is close enough to f that $2Kd(f, g) \le r$, then we have $\text{inc}_\Lambda = j\Phi(g)$ so $\Phi(g)$ must be injective. □

Corollary 8.4. *C^k Anosov diffeomorphisms, $k \ge 1$, are structurally stable and form an open subset of* $\text{Diff}^k(M)$.

PROOF. The hypothesis of the preceding theorem hold with $\Lambda = M$. It remains, then, only to show that $\Phi(g)$ is surjective. There is one proof using singular homology (see exercises) but we will give another.

We retain the notations of the preceding theorem. We begin by remarking that if g is C^1 close to f and the following diagram commutes up to α:

$$\begin{array}{ccc} M & \xrightarrow{k} & M \\ {\scriptstyle g}\big\downarrow & & \big\downarrow{\scriptstyle g} \\ M & \xrightarrow{k} & M, \end{array}$$

then the unicity in Theorem 7.8 implies that $K = \text{id}$.

Since the diagram

$$\begin{array}{ccc} M & \xrightarrow{\text{id}} & M \\ {\scriptstyle g}\downarrow & & \downarrow{\scriptstyle f} \\ M & \xrightarrow{\text{id}} & M \end{array}$$

commutes up to $d(f, g)$, there is a map $j\colon M \to M$ satisfying: $jg = fj$ and $d(\text{id}, j) \leq Kd(f, g)$. One easily sees that $d(\text{id}, j) \leq 2Kd(f, g)$ and $g \circ [\Phi(g)j] = [\Phi(g)j] \circ g$, which, when $2Kd(f, g) \leq r$, implies that $\Phi(g)j = \text{id}$ and thus that $\Phi(g)$ is surjective.

Finally, Theorem 7.6 shows that the Anosov diffeomorphisms are open. □

Proposition 8.5. *Let Λ be a closed, hyperbolic invariant set for a map $f \in \text{Diff}^k(M)$, $k \geq 1$. One can choose a neighborhood U of Λ and constants $\alpha > 0$, $K > 0$, such that if $\{x = x_0, x_1, \ldots, x_n = x\}$ is an α_1-pseudo-orbit of f contained in U with $\alpha_1 < \alpha$, then there is a periodic orbit of period n, $K\alpha_1$-close to the given pseudo-orbit:*

$$\exists x' \quad \textit{such that} \quad f^n(x') = x' \quad \textit{and} \quad d(f^i(x'), x_i) \leq K\alpha_1, \qquad i = 0, \ldots, n-1.$$

PROOF. Take U, K, and α as in Theorem 7.8. Let $X = \{x_0, x_1, \ldots, x_{n-1}\}$. Let h be the bijection $h\colon X \to X$ given by $h(x_i) = x_{i+1}$, $0 \leq i \leq n-2$, $h(x_{n-1}) = x_0$.

If i is the inclusion of X into U, the following diagram commutes up to α_1:

$$\begin{array}{ccc} X & \xrightarrow{i} & U \\ {\scriptstyle h}\downarrow & & \downarrow{\scriptstyle f} \\ X & \xrightarrow{i} & M. \end{array}$$

Thus there is a map $j\colon X \to M$ such that $fj = jh$ and $d(j, i) \leq K\alpha$. Therefore, for all positive integers k, $f^kj = jh^k$, in particular $f^kj(x_0) = jh^k(x_0) = jx_k$, $0 \leq k \leq n-1$, and $f^nj(x_0) = jh^n(x_0) = j(x_0)$. The proposition then follows, letting $x' = j(x_0)$. □

Proposition 8.6. *If $R(f)$ is hyperbolic, then $R(f) = \overline{\text{Per}(f)}$.*

PROOF. Take α and K as in Proposition 8.5. We have shown in Proposition 3.6 that $R(f|_{R(f)}) = R(f)$. Every point x of $R(f)$ is thus α_1-pseudoperiodic for all α_1 less than α, the pseudo-orbit being contained in $R(f)$.

Proposition 8.5 thus gives us, for every $\alpha_1 < \alpha$, a periodic point x' with $d(x', x) \leq K\alpha_1$, so x must belong to $\overline{\text{Per}(f)}$. □

Proposition 8.7. *If $L(f)$ (resp. $L_+(f)$, $L_-(f)$) is hyperbolic we have $\overline{\text{Per}(f)} = L(f)$ (resp. $\overline{\text{Per}(f)} = L_+(f), L_-(f)$).*

Proof. We will restrict ourselves to the case of $L_+(f)$, the other cases being similar.

First notice that, as in the proof of Proposition 1.4, if V is a neighborhood of $L_+(f)$ and y and point of M, we have

$$\exists n_0 \geq 0 \quad \text{such that} \quad \forall n \geq n_0, \quad f^n(y) \in V.$$

Choose a neighborhood U of $L_+(f)$ and constants α and K as in Proposition 8.5. Let x be a point of $L_+(f)$ and y a point of M with $x \in \omega(y)$. Now let $0 < \alpha_1 < \alpha$; the above shows that there are positive integers n and k such that

$$d(f^n(y), x) < \alpha_1/2, \qquad d(f^{n+k}(y), x) < \alpha_1/2,$$
$$f^{n+j}(y) \in U \qquad \text{for} \quad 0 \leq j \leq k,$$
$$\{f^n(y), f^{n+1}(y), \ldots, f^{n+k-1}(y), f^n(y)\} \text{ is an } \alpha_1\text{-pseudo-orbit contained in } U.$$

Proposition 8.5 then implies that there is a periodic point x' such that $d(f^n(y), x') \leq K\alpha_1$. We also have $d(x', x) \leq (K + \frac{1}{2})\alpha_1$. Since α_1 was arbitrary, x belongs to $\overline{\text{Per}(f)}$, and we are done. □

The analogous result for $\Omega(f)$ is false, cf. the commentaries. However, we can prove the following, just as we did Proposition 8.6.

Proposition 8.8. *If $\Omega(f)$ is hyperbolic, then* $\overline{\text{Per}(f)} = R(f|_{\Omega(f)})$.

Recall that in general $\Omega(f|_{\Omega(f)}) \neq \Omega(f)$, $R(f|_{\Omega(f)}) \neq \Omega(f)$. To avoid this difficulty, we make the following definition:

Definition 8.9. A map f in $\text{Diff}^r(M)$, $r \geq 1$ satisfies *axiom A* if:

(a) $\Omega(f)$ is hyperbolic, and
(b) $\Omega(f) = \overline{\text{Per}(f)}$.

Let Λ be a closed hyperbolic invariant set for f; we know that for ε sufficiently small, there is a $\delta > 0$ such that

$$x, y \in \Lambda, \qquad d(x, y) < \delta \Rightarrow W_\varepsilon^s(x) \cap W_\varepsilon^u(y) \text{ is a single point.}$$

Furthermore, this intersection is transverse. We denote this point by $[x, y]_{\varepsilon,\delta}$, suppressing the ε and δ when there is no ambiguity. The bracket map $[\cdot, \cdot]\colon U_\delta(\Delta_\Lambda) \to M$ is continuous, where $U_\delta(\Delta_\Lambda)$ is the neighborhood of the diagonal in $\Lambda \times \Lambda$ defined by $U_\delta(\Delta_\Lambda) = \{(x, y) | x, y \in \Lambda d(x, y) < \delta\}$.

Definition 8.10. A closed hyperbolic invariant set has a *local product structure* if, for small ε and δ, $[x, y]$ belongs to Λ whenever $d(x, y) < \delta$.

Proposition 8.11. *If* $\overline{\text{Per}(f)}$ *is hyperbolic, then* $\overline{\text{Per}(f)}$ *has a local product structure.*

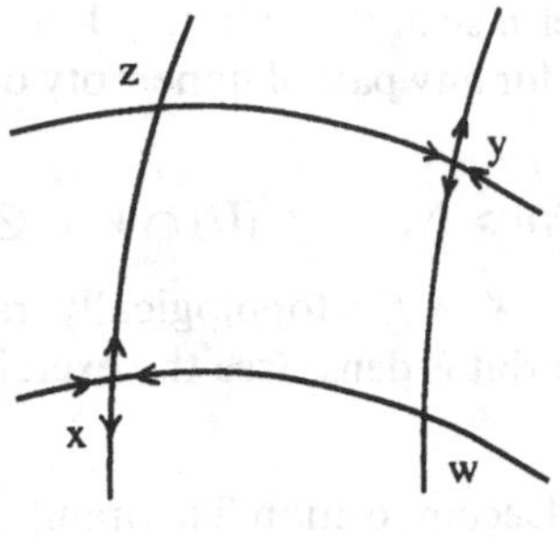

Figure 8.1.

PROOF. Take U, α, and K as in Proposition 8.5. Chose an $\varepsilon > 0$ small enough so that for all x in $\overline{\mathrm{Per}(f)}$ we have $W_\varepsilon^s(x) \cup W_\varepsilon^u(x) \subset U$. As before, we associate to ε a small positive δ. We wish to show the implication

$$d(x, y) < \delta; \qquad x, y \in \Lambda \Rightarrow [x, y] \in \Lambda.$$

Since $[\cdot, \cdot]$ is continuous, we need only consider the case when x and y are periodic.

Let $w = [x, y]$ and $z = [y, x]$ (Figure 8.1). We will show that, for all $\alpha_1 > 0$, w is α_1-pseudoperiodic and that the pseudo-orbit is contained in U. Proposition 8.5 will then show that w belongs to $\overline{\mathrm{Per}(f)}$.

Suppose that x has periodic n, and y period m. We then have

$$\forall j \geq 0, \quad f^j(w) \in W_\varepsilon^s(f^j(x)) \subset U \quad \text{and} \quad \lim_{k \to +\infty} f^{kn}(w) = x,$$

$$\forall j \geq 0, \quad f^j(z) \in W_\varepsilon^s(f^j(y)) \subset U \quad \text{and} \quad \lim_{k \to +\infty} f^{km}(z) = y,$$

$$\forall j \leq 0, \quad f^j(w) \in W_\varepsilon^u(f^j(y)) \subset U \quad \text{and} \quad \lim_{k \to -\infty} f^{km}(w) = x,$$

$$\forall j \leq 0, \quad f^j(z) \in W_\varepsilon^u(f^j(x)) \subset U \quad \text{and} \quad \lim_{k \to -\infty} f^{kn}(z) = y.$$

Choose positive integers k_1, k_2, k_3, and k_4 such that

$$d(f^{k_1 n}(w), x) < \alpha_1/2, \qquad d(f^{-k_2 n}(z), x) < \alpha_1/2,$$

$$d(f^{k_3 m}(z), y) < \alpha_1/2, \quad \text{and} \quad d(f^{-k_4 n}(w), y) < \alpha_1/2.$$

One can easily check, then, that the sequence

$$\{w, f(w), \ldots, f^{nk_1 - 1}(w), f^{-nk_2}(z), \ldots, f^{-1}(z), z, f(z), \ldots, f^{mk_3 - 1}(z),$$
$$f^{-mk_4}(w), \ldots, f^{-1}(w), w\}$$

is an α_1-pseudo-orbit contained in U. □

Definition 8.12. Let X be a compact metric space. A homeomorphism $f: X \to X$ is *topologically transitive* if, for every pair of nonempty open sets U

and V, there is an integer n such that $f^n(U) \cap V$ is nonempty. We say that f is *topologically mixing* if for any pair of nonempty open sets U and V, there is an integer N such that

$$\forall n > N, \qquad f^n(U) \cap V \neq \varnothing.$$

One can also show that $f\colon X \to X$ is topologically transitive if and only if there is a point z in X whose orbit is dense (see the exercises).

Theorem 8.13 (Spectral Decomposition Theorem). *Let f belong to* $\mathrm{Diff}^r(M)$, $r \geq 1$. *Suppose that $\overline{\mathrm{Per}(f)}$ is hyperbolic. There is a decomposition of $\overline{\mathrm{Per}(f)}$ into disjoint closed sets, $\overline{\mathrm{Per}(f)} = P_1 \cup \cdots \cup P_s$, such that*:

(a) *Each P_i is f invariant and f restricted to P_i is topologically transitive.*
(b) *There is a decomposition of each P_i into disjoint closed sets $P_i = X_{1,i} \cup \cdots \cup X_{n_i,i}$ such that $f(x_{j,i}) = X_{j+1,i}$, for $1 \leq j \leq n_i - 1$, $f(X_{n_i,i}) = X_{1,i}$, and the map $f^{n_i}\colon X_{j,i} \to X_{j,i}$ is topologically mixing for all j, $1 \leq j \leq n_i$.*

Proof. Let p be a point of $\mathrm{Per}(f)$ and δ a small positive constant. Set $X_p = \overline{W^u(p) \cap \overline{\mathrm{Per}(f)}}$ and $B_\delta(X_p) = \{y \in \overline{\mathrm{Per}(f)} \mid d(y, X_p) < \delta\}$.

First we will show that each X_p is both open and closed in $\overline{\mathrm{Per}(f)}$. Let δ be the constant in the definition of the local product structure for $\overline{\mathrm{Per}(f)}$ and y a point of $B_\delta(X_p) \cap \mathrm{Per}(f)$. Suppose the period of y is k. Let x be a point of $W^u(p) \cap \overline{\mathrm{Per}(f)}$ with $d(x, y) < \delta$; clearly, we have $W^u(p) = W^u(x)$. Set $z = [y, x]$; by Proposition 8.11, z is in $\overline{\mathrm{Per}(f)}$; by its definition z belongs to $W^s(y)$ so $f^{nk}(z)$ tends to y as n tends to infinity. On the other hand, since z belongs to $W^u(x) = W^u(p)$, if the period of p is l, $f^{nl}(z)$ belongs to $W^u(p) \cap \overline{\mathrm{Per}(f)}$, so y must belong to $\overline{W^u(p) \cap \overline{\mathrm{Per}(f)}}$, that is y is in X_p.

Since the periodic points are dense in $B_\delta(X_p)$ and $B_\delta(X_p)$ is open in $\overline{\mathrm{Per}(f)}$ we see $B_\delta(X_p) \subset X_p$. Each X_p is thus both open and closed in $\overline{\mathrm{Per}(f)}$.

Next, we show that if p and q are points of $\mathrm{Per}(f)$, X_p and X_q are either identical or disjoint. Suppose q belongs to X_p. Set $\hat{W}^u_\gamma(q) = W^u_\gamma(q) \cap \overline{\mathrm{Per}(f)}$. Since X_p is open in $\overline{\mathrm{Per}(f)}$, when γ is small enough, X_p contains $\hat{W}^u_\gamma$; thus we have

$$X_q = \overline{\bigcup_{n \geq 0} f^{nlm}(\hat{W}^u_\gamma(q))} \subset X_p,$$

where p has period l, and q has period m.

Since X_p is an open neighborhood of q in $\overline{\mathrm{Per}(f)}$, it contains a point y of $W^u(p) \cap \overline{\mathrm{Per}(f)}$, therefore we have $p = \lim_{n \to -\infty} f^{nml}(y) \in \bar{X}_q = X_q$. The previous argument then shows that $X_p \subset X_q$.

Now let p and q be arbitrary points of $\mathrm{Per}(f)$. If $X_p \cap X_q$ is not empty, it must be open so we can find a periodic point q' in the intersection. As before, then $X_p = X_{q'} = X_q$.

Since $\overline{\mathrm{Per}(f)}$ is compact there can only be a finite number of distinct X_p's; and since $X_{f(p)} = f(X_p)$, they are permuted by f.

Let $P_1, \ldots, P_s$ be the distinct f orbits of these sets

$$P_1 = X_{1,1} \cup \cdots \cup X_{n_1,l}$$

$$\vdots$$

$$P_s = X_{1,s} \cup \cdots \cup X_{n_s,s}.$$

To finish, we need only show that when $f^N(X_p) = X_p$, f^N mixes X_p, since $f^{n_i}\colon X_{j,i} \to X_{j,i}$ being mixing implies f restricted to P_i is transitive.

If U and V are arbitrary open subsets of X_p we need to find a positive integer T such that for all $t > T$, $f^{tN}(V) \cap U$ is not empty. Let p_1 be a periodic point in V. We have $X_{p_1} = X_p$ so we can find a point z in $W^u(p_1) \cap \overline{\mathrm{Per}(f)}$ such that z belongs to U. Suppose p_1 has periodic k under f^N, $f^{kN}(p_1) = p_1$. Again, for each i, $0 \le i \le k-1$, we can find a point z_i in U belonging to $W^u(f^{iN}(p_1))$. The sequence $f^{Nkt}(z_i)$ tends to $f^{iN}(p_1)$ as t tends to infinity, so we can find integers T_i, $0 \le i \le k-1$, such that

$$\forall t \ge T_i, \qquad f^{-Nkt}(z_i) \in f^{iN}(V),$$

in other words,

$$\forall t \ge T_i, \qquad f^{-Nkt-iN}(z_i) \in V.$$

Set $T = \max\{T_i\}$; when t is greater than kT, set $t = ks + i$, with $0 \le i < k$. Since s will then be greater than T_i, $f^{tN}(z_i) = f^{-Nks-iN}(z_i)$ will belong to V. Therefore,

$$\forall t \ge kT, \qquad f^{-tN}(U) \cap V \ne \varnothing,$$

in other words,

$$\forall t \ge kT, \qquad U \cap f^{tN}(V) \ne \varnothing,$$

and we are done. □

REMARK. The sets P_i and $X_{j,i}$ are unique up to indexing. Since each P_i has a dense orbit, it cannot be written as a disjoint union of a finite number of nontrivial closed in invariant sets. This implies the unicity of the P_i's, since if $Q_1, \ldots, Q_r$ were another decomposition of $\overline{\mathrm{Per}(f)}$ by disjoint and closed invariant sets, upon which f is transitive, $Q_1 \cap P_i, \ldots, Q_r \cap P_i$ would be a decomposition of P_i into disjoint closed invariant sets, and all but one of these would necessarily be empty so $P_i \subset Q_j$. Exchanging the roles of the P's and Q's we see $Q_j \subset P_i$. The unicity of the X's follows similarly.

Theorem 8.13 suggests the following definition.

Definition 8.14. Let X be a compact seperable metric space, f a continuous map of X to itself, and Y a closed invariant subset. The decomposition $Y = P_1 \cup \cdots \cup P_s$ is said to be a *spectral decomposition* of Y if:

(a) The P_i's are pairwise disjoint closed invariant sets and the restriction of f to each P_i is a topologically transitive *homeomorphism*.

(b) Each P_i is decomposed as a disjoint union of closed subsets $P_i = X_{1,i} \cup \cdots \cup X_{n_i,i}$, cyclically permuted by f such that f^{n_i} restricted to $X_{j,i}$ is topologically mixing.

The following theorem is now clear, the hyperbolic set in question in each case being equal to $\overline{\text{Per}(f)}$.

Theorem 8.15. *Let f be a C^r diffeomorphism, $r \geq 1$, of the manifold M.*

(a) *If $L(f)$ (resp. $L_+(f)$, $L_-(f)$) is hyperbolic, then $L(f)$ (resp. $L_+(f)$, $L_-(f)$) has a spectral decomposition.*
(b) *If f satisfies Axiom A, then $\Omega(f)$ has a spectral decomposition.*
(c) *If $R(f)$ is hyperbolic, then $R(f)$ has a spectral decomposition.* □

Suppose $L(f) = \overline{\text{Per}(f)}$ has a spectral decomposition $P_1 \cup \cdots \cup P_s$; we say that it has no cycles if the P_i's have no cycles (cf. Chapter 2).

Definition 8.16. Let f be a continuous map of a compact metric space to itself and suppose that Y is a closed invariant set containing $L(f)$ which admits the spectral decomposition $Y = P_1 \cup \cdots \cup P_s$. We say that it has *no cycles* if the P_i's have no cycles.

REMARK. If $R(f)$ is hyperbolic and has a spectral decomposition, it has no cycles since $R(f)$ has a sequence of filtrations and even a filtration (cf. Theorem 3.4 and Exercise 8.6).

In every case of our spectral decomposition theorems, if there are no cycles, we have a filtration **M** such that $K(\mathbf{M}) = \overline{\text{Per}(f)}$. We will show that with this hypothesis, f is L-, Ω-, or R-stable as the case may be.

Our starting point will be some observations about closed hyperbolic sets with local product structure. *A priori*, these hold in slightly more generality than setting of spectral decomposition theorem since it is not clear that the periodic points are always dense in such a set. Even the case of an Anosov diffeomorphism where all of M has local product structure is open. The strongest result our methods give is the following corollary of 8.6.

Proposition 8.17 (Anosov's Closing Lemma). *For an Anosov diffeomorphism* $\overline{\text{Per}(f)} = R(f)$. □

Suppose that Λ is a closed hyperbolic invariant set for f. Since an α_1-pseudo-orbit $\underline{x} = \{x_n\}$ can be thought of as a function i from $\mathbb{Z}$ to M with $i(n) = x_n$, Theorem 7.8 gives a neighborhood U of Λ and numbers K and α such that if $\alpha_1 < \alpha$ and $\underline{x}$ is contained in U, then we can find a map $j\colon \mathbb{Z} \to M$ such that $d(i, j) \leq K\alpha_1$ and $j(n+1) = f(j(n))$, in other words, letting $y_n = j(n)$, we have found an orbit $y_n = f(y_0)$ satisfying $d(y_n, X_n) \leq K\alpha_1$, for all n. With this in mind we make the following definition.

Definition 8.18. Let f be a map of the metric space X to itself. Let $\underline{x} = \{x_i | a < i < b\}$ be an α-pseudo-orbit of f. A point x' *β-shadows* $\underline{x}$ if $\forall i$, $a < i < b$ $d(f^i(x'), x_i) \leq \beta$.

The preceding discussion can now be restated as:

Proposition 8.19. *If Λ is a closed hyperbolic invariant set, there is a neighborhood U of Λ and positive constants α and K such that every α_1-pseudo-orbit $\underline{x}$ with $\alpha_1 < \alpha$ contained in U is $K\alpha_1$ shadowed by a point x' in M.* □

We would like to be able to assert that x' is in Λ when $\underline{x}$ is; in order to do this we need to assume that Λ has a local product structure.

Proposition 8.20 (Shadowing Lemma). *Let Λ be a closed hyperbolic invariant set for a C^r diffeomorphism of a manifold M. Suppose that Λ has a local product structure. Then for every $\beta > 0$ there is an $\alpha > 0$ such that every α-pseudo-orbit $\underline{x}$ in Λ is β-shadowed by a point y of Λ.*

Proof. Suppose M has an adapted metric. Choose an ε as in the stable manifold theorem for Λ, and let $\lambda \in (0, 1)$ be the constant of hyperbolicity of Λ. Now choose a positive ε_1, less than $(1 - \lambda)\min\{\varepsilon, \beta\}$. Let $\eta = \varepsilon_1/(1 - \lambda)$ and let δ be a positive constant less than $\beta - \eta$ for which $[\cdot, \cdot]_{\varepsilon_1, \delta}: U_\delta(\Delta_\Lambda) \to \Lambda$ defines a local product structure.

Since $[\cdot, \cdot]$ is continuous, as is $W^s_{loc}(\cdot)$, it makes sense to define α by requiring that whenever z and w in Λ are α close

$$[z, W^s_{\lambda\delta}(w) \cap \Lambda] \subset W^s_\delta(z).$$

First suppose that the pseudo-orbit $\underline{x}$ has the special form $\underline{x} = \{x_0, \ldots, x_n\}$. Set $y_0 = x_0$ and define y_k recursively by $y_k = [x_k, f(y_{k-1})]$ for $1 \leq k \leq n$; in order for this to be possible we need y_k to belong to $W^s_\delta(x_k) \cap \Lambda$. Suppose by induction that y_{k-1} belongs to $W^s_\delta(x_{k-1}) \cap \Lambda$. Then $f(y_{k-1})$ belongs to $W^s_{\lambda\delta}(f(x_{k-1}))$ and hence to $W^s_\delta(x_k)$ as well, so our definition is valid. Note that y_k belongs to $W^u_{\varepsilon_1}(f(y_{k-1}))$ by our definition of this local product structure, so, by descent, $f^{-j}(y_k)$ belongs to $W^u_{\theta_j}(y_{k-j})$, where $\theta_j = \sum_{i=1}^{j} \lambda^i \varepsilon_1 \leq \gamma = \varepsilon_1/(1 - \lambda)$. We claim the point $y = f^{-n}(y_n)$ β-shadows $\underline{x}$. Since $f^{-(n-j)}(y_n) = f^j(y)$ belongs to $W^u_\eta(y_j)$, we have

$$d(f^j(y), x_j) \leq d(f^j(y), y_j) + d(y_j, x_j) \leq \eta + \delta < \beta.$$

The case of an arbitrary finite pseudo-orbit proceeds as above, after reindexing.

Finally, if $\underline{x}$ is infinite, we can find points $\hat{y}$ in Λ which β-shadow each finite segment $\hat{\underline{x}}$ of $\underline{x}$, and since Λ is compact the $\hat{y}$'s have a limit y which β-shadows $\underline{x}$. □

We can refine the previous result, using expansivity.

Proposition 8.21. *Let* ε *be a constant of expansivity of* f *on* Λ, f *and* Λ *as in Proposition* 8.20, *and let* γ *be a positive constant less than* $\varepsilon/2$. *Then*

(a) *A bi-infinite pseudo-orbit* $\underline{x}$ *is* γ*-shadowed by at most one point* y *in* Λ.
(b) *There is a constant* α *and a neighborhood* U *of* Λ *such that every* α*-pseudo-orbit* $\underline{x}$ *in* U *is* γ *shadowed by a point* y *in* Λ. *If, moreover,* $\underline{x}$ *is bi-infinite, then* y *is unique.*

Proof. (a) If y_1 and y_2 γ-shadow $\underline{x}$, we have

$$d[f^n(y_1), f^n(y_2)] \leq 2\gamma < \varepsilon, \qquad \forall n \in \mathbb{Z},$$

so y_1 and y_2 must coincide.

(b) Let α_1 correspond to the choice $\beta = \gamma/2$ in Proposition 8.20. Choose a neighborhood U of Λ and a constant α such that every pseudo-orbit $\underline{x}$ in U is approximated to within $\gamma/2$ by an α_1-pseudo-orbit $\underline{x}'$ in $\Lambda(d(x_i, x_i') < \gamma/2, \forall i)$; this is possible since f is uniformly continuous. The pseudo-orbit $\underline{x}'$ is $\gamma/2$-shadowed by a point x of Λ, which must also γ-shadow $\underline{x}$. Unicity follows from (a). □

Proposition 8.22. *Let* Λ *be a closed hyperbolic invariant set for* f *which has a local product structure.* Λ *is uniformly locally maximal. More precisely, there are neighborhoods* $U \subset M$ *of* Λ *and* $V \subset \mathrm{Diff}^r(M)$ *of* f *such that*:

(1) $\Lambda = \bigcap_{n \in \mathbb{Z}} f^n(U)$.
(2) *The set* $\Phi(g)\Lambda$ *given by Theorem* 8.3 *for* $g \in V$, *conjugate to* Λ, *is equal to* $\bigcap_{n \in \mathbb{Z}} g^n(U)$.

Proof. Take U small enough so that Proposition 8.21, Theorem 7.8 apply. V will be a very small neighborhood of f where the map Φ of Theorem 8.3 is defined and, furthermore, for all g in V, $\sup_{x \in M} d(g(x), f(x)) < \alpha$, where α is as in Proposition 8.21. Whenever z belongs to $\bigcap_{n \in \mathbb{Z}} g^n(U)$, the g-orbit of z will also be a bi-infinite α-pseudo-orbit of f, contained in U. Thus it will be γ-shadowed by a unique point x of Λ.

We claim that $\Phi(g)x = z$. First notice that by taking a small enough V, we can guarantee that

$$d(f^n(x), g^n(\Phi(g)x)) = d(f^n(x), \Phi(g)f^n(x)) \leq d(\mathrm{id}, \Phi(g)) < \delta.$$

Next, considering the diagram

$$\begin{array}{ccc} \mathbb{Z} & \xrightarrow{i} & \Lambda \\ {\scriptstyle h}\downarrow & & \downarrow{\scriptstyle g} \\ \mathbb{Z} & \xrightarrow{i} & \Lambda, \end{array}$$

where $i(n) = f^n(x)$ and $h(n) = n + 1$, Theorem 7.8 tells us, again for δ and hence V sufficiently small, that there is a unique point y such that

$d(f^n(x), g^n(y)) < \delta$, for all n. Since both z and $\Phi(g)x$ satisfy this condition, they must be equal.

We have shown, then, that $\bigcap_{n \in \mathbb{Z}} g^n(U) \subset \Phi(g)\Lambda$. The continuity of Φ, on the other hand, allows us to find a small enough V so that whenever g is in V, $\Phi(g)\Lambda$ will be contained in U and hence also in $\bigcap_{n \in \mathbb{Z}} g^n(U)$. □

Theorem 8.23. *Suppose that Λ is a closed hyperbolic invariant set for f which has a local product structure and contains $L(f)$. Suppose, further, that Λ has a decomposition $\Lambda = \Lambda_1 \cup \cdots \cup \Lambda_k$ by disjoint closed invariant sets and the Λ_i's have no cycles. Then there is a filtration $\mathbf{M}$ adapted to Λ, a neighborhood V of f in* $\text{Diff}^r(M)$, *and a continuous function $\Phi: V \to C^0(\Lambda, M)$ such that:*

(1) $\mathbf{M}$ *is adapted simultaneously to all maps g in V.*
(2) $\Phi(f) = \text{inc}_\Lambda$.
(3) $\Phi(g)\Lambda = K^g(\mathbf{M})$; $\Phi(g)\Lambda_i = K_i^g(\mathbf{M})$.
(4) $\Phi(g): \Lambda \to K^g(\mathbf{M})$ *is a topological conjugacy.*
(5) *There is a positive constant K such that Φ is Lipschitz at f with respect to the C^0 metric; that is $d(\Phi(g), \text{inc}_\Lambda) \leq Kd(g, f)$.*

Proof. The existence of $\mathbf{M}$ is given by Theorem 2.4 and (1) is a consequence of Proposition 2.10. Proposition 2.10 also allows us to see that for g close enough to f, $K^g(\mathbf{M})$ will be contained in the neighborhood U of Λ in Proposition 8.20. Consequently, $K^g(\mathbf{M})$ is contained in $\Phi(g)\Lambda$, and we will be done if we can demonstrate the opposite inclusion. Writing the filtration $\mathbf{M}$ as $\varnothing \neq M_0 \subset M_1 \subset \cdots \subset M_k = M$, we have each closed invariant set Λ_i contained in the difference $M_i - \overline{M_{i-1}}$. If g is close enough to f, then $\Phi(g)\Lambda_i$ will also be in $M_i - \overline{M_{i-1}}$ so $\Phi(g)\Lambda_i$ is contained in $K_i^g(\mathbf{M})$. □

This proposition has the following corollary, which we reformulate three times for dramatic effect.

Corollary 8.24 (Ω-stability Theorem). *Let f be a C^r diffeomorphism of the manifold M.*

(a) *If $L(f)$ is a hyperbolic set for f and has no cycles, then f is L-stable. The set of such f is open in* $\text{Diff}^r(M)$.
(b) *If f satisfies Axiom A and has no cycles, then f is Ω-stable. The set of such f is open in* $\text{Diff}^r(M)$.
(c) *If $R(f)$ is hyperbolic, then f is R-stable. The set of such f is open in* $\text{Diff}^r(M)$.

In fact, in each of the cases (a), (b), and (c) we have $\overline{\text{Per}(f)} = L(f) = \Omega(f) = R(f)$.

Proof. We have seen the equality of $\overline{\text{Per}(f)}$, $L(f)$, $\Omega(f)$, and $R(f)$ in cases (a), (b), and (c) in 8.6–8.9. We have remarked upon the lack of cycles in the spectral decomposition of a hyperbolic $R(f)$ after Definition 8.16 and we have shown in Proposition 8.11 that $\overline{\mathrm{Per}(f)}$ has a local product structure when it is hyperbolic. Applying Theorem 8.23 to $\Lambda = \overline{\mathrm{Per}(f)}$, we see that when g is in a certain neighborhood V of f we have $\Phi(g)\Lambda = K^g(\mathbf{M})$, hence

$$\overline{\mathrm{Per}(g)} \subset L(g) \subset \Omega(g) \subset R(g) \subset \Phi(g)\Lambda.$$

Since $\Phi(g)$ is a conjugacy, $\Phi(g)\,\mathrm{Per}(f)$ is contained in $\mathrm{Per}(g)$ and $\Phi(g)\,\overline{\mathrm{Per}(f)} = \Phi(g)\Lambda$ is contained in $\overline{\mathrm{Per}(g)}$ so we actually have equality

$$\overline{\mathrm{Per}(g)} = L(g) = \Omega(g) = R(g) = \Phi(g)\Lambda.$$

When g is close to f, Theorem 8.3 tells us that $\Phi(g)\Lambda$ is hyperbolic for g. Finally, we see that the spectral decomposition of $g|_{\Phi(g)\Lambda}$ has no cycles since Theorem 8.23 gives us an adapted filtration $\mathbf{M}$. □

REMARK. Along the way we have shown that the three conditions, $L(f)$ is hyperbolic with no cycles, f satisfies Axiom A and has no cycles, and $R(f)$ is hyperbolic, are all equivalent. In other words, in all known cases of L-, Ω-, and R-stability the three results coincide.

CENTRAL PROBLEM. Does Ω (resp. L, R)-stability of a C^r diffeomorphism f imply that $\Omega(f)$ (resp. $L(f)$, $R(f)$) is hyperbolic?

Traditionally this question is posed only for Ω. The answer is only known for the circle. (See commentaries.)

PROBLEM. If f is Anosov, does $\overline{\mathrm{Per}(f)} = M$, or, equivalently, does $R(f) = M$?

EXERCISE 8.1. Show that if a homeomorphism f of a compact metric space X is topologically transitive, then there is a point z of X whose orbit is dense.

EXERCISE 8.2. Formulate and prove an L_+ stability theorem using the concepts of cycle and filtration developed in the exercises of Chapter 3.

EXERCISE 8.3. Let Λ be a closed hyperbolic invariant set for $f \in \mathrm{Diff}^r(M)$ with a local product structure. Show, generalizing Proposition 8.19, that $\overline{\mathrm{Per}(f|_\Lambda)} = R(f|_\Lambda)$. Show that an Anosov diffeomorphism satisfies axiom A and has no cycles.

EXERCISE 8.4. Let f be an Anosov diffeomorphism of the manifold M. Show that there is a neighborhood V of f in Homo(M) such that for every map g in V there is a semiconjugacy from f to g, that is continuous, surjective map $h: M \to M$ such that $hg = fh$. *Hint*: To show that h is surjective note that a continuous map j of M to itself which is C^0 close to the identity is necessarily homotopic to the identity. Therefore, j induces the identity map on top dimensional homology with $\mathbb{Z}_2$ coefficient

$$j_* = \mathrm{id}: H_m(M, \mathbb{Z}_2) \to H_m(M, \mathbb{Z}_2) = \mathbb{Z}_2,$$

and we see that j cannot factor through M-pt. since $H_m(M\text{-pt.}) = 0$. In other words j is surjective.

EXERCISE 8.5. If f is an Anosov diffeomorphism of M, show that there is a neighborhood V of the identity in $C^0(M, M)$ such that the only map j in V such that $fj = jf$ is the identity.

EXERCISE 8.6. Let $P_1 \cup \cdots \cup P_s$ be the spectral decomposition of $f|_{\overline{\mathrm{Per}(f)}}$ given in Theorem 8.13. Show that not only is the homeomorphism $f|_{P_i}$ topologically transitive, but also if U and V are nonempty open sets and n a positive integer, then there is an $m > n$ such that $f^m(U) \cap V$ is not empty. *Hint*: Use the fact that $f^{n_i}|_{X_{j,i}}$ is topologically mixing.

EXERCISE 8.7. Show that if $R(f)$ is hyperbolic, then it has no cycles. *Hint*: Suppose that $P_1, \ldots, P_r$ is a cycle. Take $x_i \in (W^u(P_i) - P_i) \cap (W^s(P_{i+1}) - P_{i+1})$, $1 \le i < r$ and $x_r \in (W^u(P_r) - P_r) \cap (W^s(P_1) - P_1)$. Show that there is a filtration M: $\varnothing = M_0 \subset M_1 \subset \cdots \subset M_k = M$ such that $x_i \notin K(\mathbf{M})$ for all i. Then show that for each i there is an index $\alpha(i)$ such that $P_i \subset K_{\alpha(i)}(\mathbf{M})$. Hence we have that $K_{\alpha(i)}(\mathbf{M}), \ldots, K_{\alpha(r)}(\mathbf{M})$ form a cycle, which is absurd.

Another hint: Suppose that $P_1, \ldots, P_r$ is a cycle, and choose $x_1, \ldots, x_r$ as before. Show that x_1 belongs to $R(f)$ by constructing an α-pseudo-orbit of the form:

$$\{x_1, f(x_1), \ldots, f^{n_1}(x_1), z_2, f(z_2), \ldots, f^{l_2}(z_2),$$
$$f^{-m_2}(x_2), \ldots, x_2, \ldots, f^{n_2}(x_2), z_3, \ldots, f^{l_r}(z_r), f^{-m_r}(x_r), \ldots, f^{n_r}(x_r), z_1, \ldots,$$
$$f^{l_1}(z_1), f^{-m_1}(x_1), \ldots, x_1\},$$

where $z_i \in P_i$.

Yet another hint: Since $R(f)$ has a sequence of filtrations, use Proposition 8.22 to construct a filtration $\mathbf{M}$ with $K(\mathbf{M}) = R(f)$. In other words, since we have a sequence of filtrations M_i with $\bigcap_i K(\mathbf{M}_i) = R(f)$, eventually one of the closed sets $K(\mathbf{M}_i)$ will be contained in the neighborhood U of $R(f)$ supplied by Proposition 8.22; $\mathbf{M}_i$ is the filtration we seek.

EXERCISE 8.8. Prove the Birkhoff–Smale Theorem. If p is a hyperbolic fixed point of diffeomorphism f and q is a point of transversal intersection of $W^s(p)$ and $W^u(p)$, then q belongs to $\overline{\mathrm{Per}(f)}$ and f has infinitely many periodic points. *Hint*: Use Proposition 8.19 and the sketch in the commentaries.

Commentary

In this chapter we reap the harvest of our labors. Once again I have taken from [1.16] the discussion of stability. Theorem 8.3 is found in almost exactly the same form in [1.16] and is formulated in [4.2]. A historical account of this subject will be found in the commentaries on Chapter 6. Corollary 8.3 is a theorem of Anosov [4.1]. Proposition 8.6 is formulated in [8.2]. Proposition 8.7 is proved in [2.1]. Allan Dankner has constructed a hyperbolic diffeomorphism with Per $f \neq \Omega(f)$ [8.5].

Proposition 8.8 is sometimes known as Anosov's closing lemma, particularly in the case where $\Omega = M$ and f is an Anosov diffeomorphism. Definition 8.9 comes from [1.16]. The bracket $[\cdot, \cdot]$ is defined in [1.16] as is the local product structure.

The notion of spectral decomposition also comes from [1.16]. Theorem 8.13 is formulated for Ω and Axiom A diffeomorphisms, [2.1] generalizes this result to $\overline{\mathrm{Per}\, f}$ when this set is hyperbolic, and thus to $L(f)$; the case of $R(f)$ is treated in [8.1].

The part concerning topologically mixing maps was treated by Bowen. The formulations and proofs are found in [1.2]. The notion of β-shadowing comes from [1.2] as does Proposition 8.20. Proposition 8.21 is formulated in [8.2], but one will find a greatly improved proof in [1.2]. We see in Theorem 8.23 the fundamental idea of Ω-stability, which comes from [1.16]. The hyperbolic structure guarantees local stability, while the filtration controls the global behavior of the diffeomorphism.

The stability theorem comes from [1.16]. The generalization to $L(f)$ is found in [2.1], and a formulation in terms of $R(f)$ in [8.1].

The central problem is certainly central. In the case of Ω it is a conjecture of Smale (see [8.3]). A good reference for what was known in this direction in 1974 is [8.4]; for more recent results and the two-dimensional C^1 theorem, see [8.6], [8.7], and [8.9]. Mañé has also proven a three-dimensional theorem. This spring Mañé [8.8] has shown that even ξ-stability implies hyperbolicity of $R(f)$ in the C^1 diffeomorphism of arbitrary compact M. All $r > 1$ remain open.

If f is a C^r diffeomorphism of M which satisfies Axiom A and has no cycles, then, when g is sufficiently close to f, the restrictions $f|_{\Omega(f)}$ and $g|_{\Omega(g)}$ are topologically conjugate via $\Phi(g)$. Theorem 8.23 shows that $\Phi(g)$ satisfies

$$d_{C^0}(\mathrm{inc}(\Omega(f)), \Phi(g)) \leq K d_{C^0}(f, g),$$

where K is a positive constant.

Conversely, if, for each g sufficiently close to f, there is a conjugacy $\Phi(g)$ from $f|_{\Omega(f)}$ to $g|_{\Omega(g)}$ satisfying this last condition, then Mañé [8.4] shows that f satisfies Axiom A.

We have, then, an excellent necessary and sufficient condition for this sort of Ω-stability, which is sometimes known as absolute stability. Some would be content with this result, but recently Mañé has studied the more difficult problem of characterizing structural stability [8.7], a subject to which we will return in the commentaries to Chapter 9.

A more detailed analysis of transversally homoclinic points shows that when $\overline{\mathrm{Per}(f)}$ is hyperbolic, this set has a local product structure.

Recall that we call a point x transversely homoclinic if it is a point of transverse intersection of the stable and unstable manifold, $W^s(\rho)$ and $W^u(\rho)$ of a hyperbolic periodic point $\rho \neq x$. The point x then belongs to $\overline{\mathrm{Per}(f)}$ [4.6]. However, since x is not periodic, $\mathrm{Per}(f)$ must be infinite.

The result is known as the Birkhoff–Smale Theorem, even though an analysis of its proof shows that it is just one aspect of a more profound result.

Suppose that ρ is a fixed point. We may schematize the situation as shown in Figure C8.1. We take a very narrow rectangle "parallel" to $W^s(\rho)$ which contains both x and ρ. Up to replacing f be a high iterate f^n, if the rectangle is narrow enough, the image of the rectangle looks as shown in Figure C8.2.

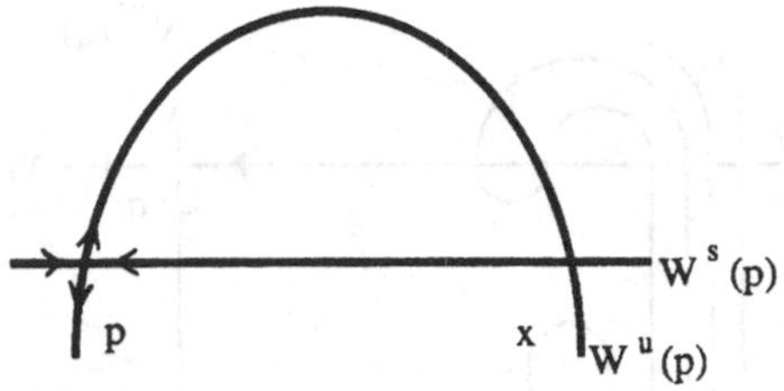

Figure C8.1.

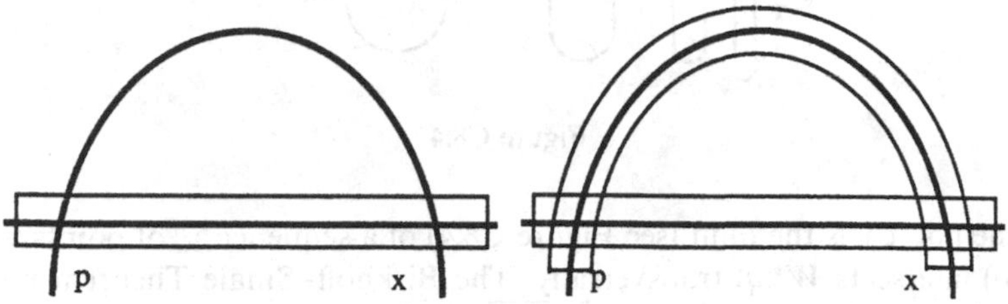

Figure C8.2.

We recover, then, the picture of the horseshoe. The set $\{f^n(x) | n \in \mathbb{Z}\} \cup \{\rho\}$ is hyperbolic for f.

To see this we take

$$E^s_x = T_x W^s(\rho), \qquad E^u_x = T_x W^u(\rho);$$

$$E^s_{f^n(x)} = Df^n T_x W^s(\rho), \qquad E^u_{f^n(x)} = Df^n T_x W^u(\rho);$$

split $T_\rho M$ as $T_\rho M = E^s_\rho \oplus E^u_\rho$, then estimate the constants of hyperbolicity. Shadowing now permits us to show directly that x belongs to $\overline{\mathrm{Per}(f)}$.

In fact, it is always possible to show that a sufficiently high iterate of f contains a 2-shift [4.6].

Now suppose that ρ and q are two periodic points and that $W^s(\rho)$ and $W^u(q)$ intersect transversally at the point x, while $W^s(q)$ and $W^u(\rho)$ intersect transversally at y (Figure C8.3). Restrict f to $W^u(q)$.

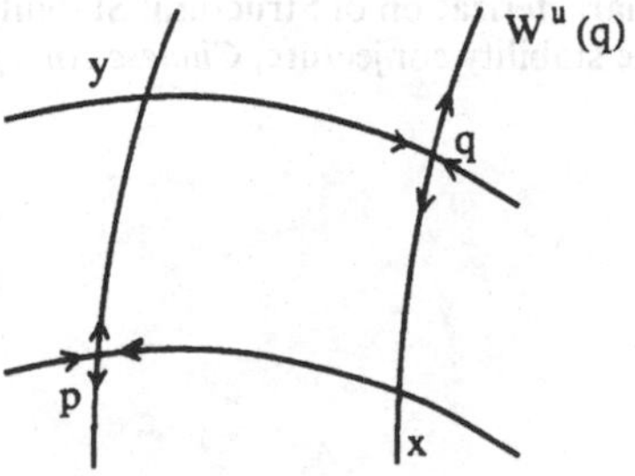

Figure C8.3.

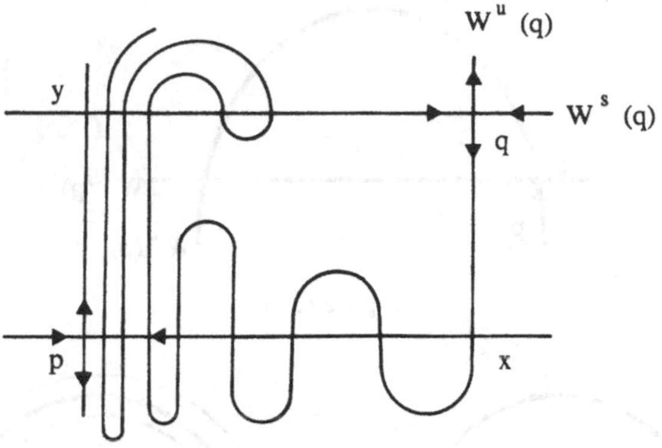

Figure C8.4.

The point y is the limit (see Figure C8.4) of a sequence y_n of points where $W^u(q)$ intersects $W^s(q)$ transversally. The Birkhoff–Smale Theorem implies that all the points y_n belong to $\overline{\mathrm{Per}(f)}$, so the point y must also belong to $\overline{\mathrm{Per}(f)}$. This result implies that $\overline{\mathrm{Per}(f)}$ has a local product structure, if it is hyperbolic; it is stronger, however, since ρ and q are not necessarily close.

References

[8.1] Franke, J. and Selgrade, J., Hyperbolicity and chain recurrence, *Trans. Amer. Math. Soc.* **245** (1978), 251.

[8.2] Hirsch, M. W., Palis, J., Pugh, C. C. and Shub, M., Neighborhoods of hyperbolic sets, *Invent. Math.* **9** (1969–70), 121.

[8.3] Smale, S., Notes on differentiable dynamical systems, in *Global Analysis*, p. 277.

[8.4] Mañé, R., On infinitesimal and absolute stability of diffeomorphisms, in *Dynamical Systems—Warwick* 1974, Manning (Ed.), Lecture Notes in Mathematics, No. 468, Springer-Verlag, New York, 1975, p. 151.

[8.5] Dankner, A., On Smale's Axiom A dynamical systems, *Ann. of Math.* **107** (1978), 517.

[8.6] Mañé, R., Contributions to the stability conjecture, *Topology* **17** (1978).

[8.7] Mañé, R., An ergodic closing lemma, *Ann. of Math.* **116** (1982).

[8.8] Mañé, R., The Characterization of Structural Stability, Preprint 5/86.

[8.9] Liao, S. T., On the stability conjecture, *Chinese Ann. Math.* **1** (1980), 19.

CHAPTER 9

A Potpourri of Stability Results

Proposition 9.1. *Let Λ be a closed invariant hyperbolic set for $f \in \mathrm{Diff}^r(M)$ which has a local product structure. We now have*

$$W^s(\Lambda) = \bigcup_{x \in \Lambda} W^s(x) \quad \textit{and} \quad W^u(\Lambda) = \bigcup_{x \in \Lambda} W^u(x).$$

PROOF. As in the *shadowing lemma* (Proposition 8.21), if we are given a sufficiently small δ, we can find a neighborhood U of Λ and a $\alpha > 0$ such that every α-pseudo-orbit of f in U is δ-shadowed by a point of Λ. For an arbitrary y in $W^s(\Lambda)$ there is an N so large that

$$\forall n \geq N, \qquad f^n(y) \in U.$$

Now the set $\underline{y} = \{y_i \mid y_i = f^{i+N}(y),\ i \geq 0\}$ is an orbit in U, and is therefore δ-shadowed by some x in Λ, that is

$$\forall i \geq 0, \qquad d(f^i(x), f^{i+n}(y)) < \delta.$$

If δ is small enough so that the local stable manifold $W^s_\delta(x)$ is defined, $f^N(y)$ must belong to it, so y belongs to the stable manifold $W^s(f^{-N}(x))$.

As usual the result for W^u follows as above, by considering f^{-1}. □

Recall that if y belongs to $W^s(x)$, the stable manifolds $W^s(x)$ and $W^s(y)$ coincide. In fact, for every z in M, the two following conditions are equivalent:

(1) $$\lim_{n \to +\infty} d[f^n(z), f^n(y)] = 0,$$

(2) $$\lim_{n \to +\infty} d[f^n(z), f^n(x)] = 0,$$

since $\lim_{n \to +\infty} d[f^n(y), f^n(x)] = 0$.

If $L(f)$ is hyperbolic, $L(f)$ has a spectral decomposition and local product structure. Let $L(f) = L_1 \cup \cdots \cup L_p$ be the spectral decomposition. For every x in M, there is an index i such that x belongs to $W^s(L_i)$ and an index j such that x belongs to $W^u(L_j)$. Proposition 9.1 guarantees the existence of points y_i in L_i and y_j in L_j such that x belongs to $W^s(y_i)$ and $W^s(y_j)$. Thus one has $W^s(x) = W^s(y_i)$ and $W^u(x) = W^u(y_j)$ so we have proved:

Corollary 9.2. *If $L(f)$ is hyperbolic for $f \in \mathrm{Diff}^r(M)$, then for any point x in M, the subsets $W^s(x)$ and $W^u(x)$ are Euclidean spaces immersed in M by an injective C^r immersion.* □

The preceding result, of course, also holds when f satisfies Axiom A or $R(f)$ is hyperbolic.

Definition 9.3. If either $L(f)$ is hyperbolic, or f satisfies Axiom A, or $R(f)$ is hyperbolic, we say that f *satisfies the strong transversality condition* when, for every x in M, the stable and unstable manifolds $W^s(x)$ and $W^u(x)$ are transverse at x.

For historical reasons we chose to state the following theorem, which we will not prove, for a map f satisfying Axiom A.

Theorem 9.4. *If a C^r $(r \geq 1)$ diffeomorphism f of a compact manifold M satisfies Axiom A and the strong transversality conditions, then f is structurally stable.*

The converse is a central open problem: *If a C^r diffeomorphism f is structurally stable, must it satisfy Axiom A and the strong transversality condition?*

Definition 9.5. We denote by $\mathrm{AS}^r(M)$ the set of C^r diffeomorphisms satisfying Axiom A and the strong transversality condition. If f belongs to $\mathrm{AS}^r(M)$ and $\Omega(f)$ is finite, we say that f is *Morse–Smale*. We denote by $\mathrm{MS}^r(M)$ the set of such diffeomorphisms.

We will now outline several propositions in order ultimately to prove that an Axiom A diffeomorphism has no 1 cycles.

Proposition 9.6. *Suppose that p is a hyperbolic periodic point for a C^r diffeomorphism f of M. Suppose that the manifold W intersects $W^s(p)$ transversally at a point x in its interior, and that the manifold V intersects $W^u(p)$ transversally at a point y in its interior. Then there is a positive integer n such that $f^n(W)$ and V intersect transversally at a point z* (see Figure 9.1).

Sketch of Proof. By replacing f by one of its iterates we may assume that p is fixed. Since $W^s(p)$ is invariant we can replace W by $f^n(W)$ and x by $f^n(x)$, for any positive n, so that we may also assume that x belongs to a small

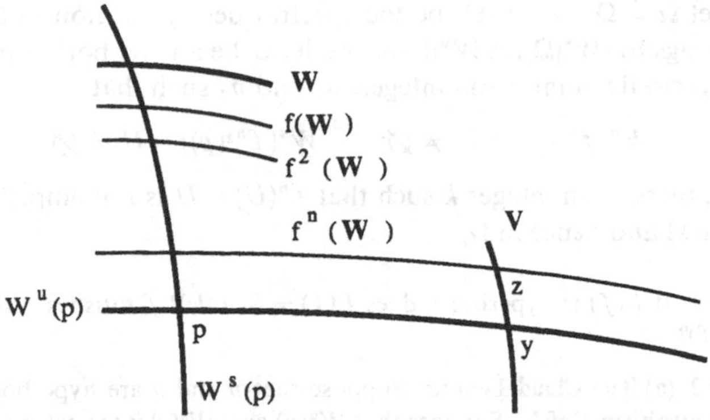

Figure 9.1.

neighborhood U of p. Replacing y by $f^{-m}(y)$ for some positive m, we nay also assume that y belongs to U. By refining the estimates we made about the graph transform, one can show that, when the dimension of W is equal to that of the manifold $W^u_{loc}(p)$, as n tends to infinity the plaque $f^n(W) \cap U$ tends toward the plaque $W^u_{loc}(p)$ in the C^1 topology, so that for n large enough $f^n(W)$ will intersect V transversally. If the dimension of W is greater than that of $W^u_{loc}(p)$, we can choose a submanifold W' of W of the same dimension as $W^u_{loc}(p)$, transverse to $W^s(p)$ at x, and proceed as above. □

REMARKS. (1) One can apply Proposition 9.6 when V and W are open subsets of M, since an open subset meets any submanifold transversally. It is also easy to give a direct proof of the proposition in this case, using our earlier techniques.

(2) If V_x is an open neighborhood in M of a point x in $W^u(p)$ and W is transverse to $W^s(p)$, then there is a positive integer n such that $f^n(W) \cap V_x$ is nonempty. Thus we have the inclusion

$$W^u(p) \subset \overline{\bigcup_{n>0} f^n(W)}.$$

Proposition 9.7. *Suppose that f satisfies Axiom A or that $L(f)$ is a hyperbolic invariant set. Let $P_1 \cup \cdots \cup P_s = \overline{\mathrm{Per}(f)}$ be the spectral decomposition. Let U and V be two nonempty open sets in P_i. Then there is a periodic point p and integers n_1 and n_2 such that $f^{n_1}(p)$ belongs to U and $f^{n_2}(p)$ belongs to V.*

PROOF. The restriction of f to P_i has a dense orbit, so

$$\exists x, n_1, n_2 \quad \text{such that} \quad f^{n_1}(x) \in U, \quad f^{n_2}(x) \in V.$$

Since the periodic points are dense, choosing a periodic point p very close to x finishes the proof. □

Proposition 9.8. *A diffeomorphism satisfying Axiom A has no 1-cycles.*

PROOF. Let $\Omega = \Omega_1 \cup \cdots \cup \Omega_s$ be the spectral decomposition of Ω. Suppose that x belongs to $(W^s(\Omega_i) \cap W^u(\Omega_i))$, and let U be a neighborhood of x. Then there is a periodic point p and integers n_1 and n_2 such that

$$W^u(f^{n_1}(p)) \cap U \neq \emptyset; \qquad W^s(f^{n_2}(p)) \cap U \neq \emptyset.$$

Therefore, there is an integer k such that $f^n(U) \cap U$ is not empty. The point x is thus in Ω and hence in Ω_i. □

EXERCISE 9.1. If $L(f)$ is hyperbolic, does $L(f) = R(f)$? If f satisfies Axiom A, does $\Omega(f) = R(f)$?

EXERCISE 9.2. (a) **The Cloud Lemma.** Suppose that p and q are hyperbolic points of a C^r diffeomorphism f of M. Suppose that $W^u(p)$ and $W^s(q)$ intersect transversally at x and that $W^u(q)$ and $W^s(p)$ intersect at y. Show that y belongs to $\Omega(f)$. (Draw a picture in order to understand the name.)

(b) Extend the preceding to a sequence of hyperbolic points of f. More precisely, if $p_1, \ldots, p_k$ are hyperbolic points of f, if $W^s(p_1)$ and $W^u(p_k)$ intersect at x, and if $W^u(p_i)$ and $W^s(p_{i+1})$ have a point of transverse intersection for $1 \leq i < k - 1$, show that x is nonwandering.

EXERCISE 9.3. Let f be a diffeomorphism satisfying Axiom A. We say that f satisfies Axiom B if, moreover, whenever $[W^s(\Omega_i) - \Omega_i] \cap [W^u(\Omega_j) - \Omega_j]$ is nonempty, there are periodic points p in Ω_i and q in Ω_j such that $W^s(p)$ and $W^u(q)$ have a point of transverse intersection.

Show that Axiom B implies no cycles.

Show that strong transversality implies Axiom B. (Thus strong transversality implies no cycles.)

EXERCISE 9.4. If $L(f)$ is hyperbolic, can $L(f)$ have a 1-cycle?

Commentary

This chapter is the weakest one of the course, since the goal which I originally set was practically attained in Chapter 8. That which remained, however, was too important to be relegated to the commentaries.

Proposition 9.1 was first proved in [8.2] in response to a question of Smale; the proof that I give comes from [1.2] and is much simpler than the original. The strong transversality condition was first formulated in [1.16] and is found in the form I give in [9.3].

Particular cases of Proposition 9.4 were proved by Peixoto for S^1, as a special case of this theorem on vector fields on surfaces [1.10], by Anosov for Anosov diffeomorphisms [4.1], by Palis for Morse–Smale diffeomorphisms in dimensions 2 and 3 [2.3], by Palis and Smale for Morse–Smale diffeomorphisms in arbitrary dimension [6.3], by Robbin, using a functional approach for f of class C^2 [6.4], and by Robinson for $C^1 f$ [6.5]. Since then numerous other treatments of this theorem, both functional and geometrical, have been given.

The essential open problem in this chapter is the conjecture of Palis–Smale [6.3] that Axiom A and the strong transversality condition are necessary and sufficient for structural stability.

It is known that a structurally stable diffeomorphism which satisfies Axiom A must also satisfy the strong transversality condition [9.2]. The key, then, is to find out if a structurally stable diffeomorphism satisfies Axiom A. We have already encountered the problem—does Ω-stability imply Axiom A?—in the preceding chapter. I cannot imagine that the answers to these two questions could be different.

Recently, Mañé [8.7] has proved the Palis–Smale stability conjecture for surfaces.

With regard to Ω-stability, it is known that Ω-stability and Axiom A imply that there are no cycles. It is this result which brought the no-cycle condition to the subject as a substitute for Smale's Axiom B [1.16] (see Exercise 9.3).

In our list of abbreviations AS, KS, MS, AS stands for Axiom A + strong transversality, but one could also read it as an abbreviation for Anosov–Smale, which seems like a good idea to me.

The convergence toward $W^u(\rho)$ that we use in Proposition 9.6 was proved in [4.2]. In order to make our sketch rigorous, we need Lemma A of [2.3].

Proposition 9.8 is there to show that Smale's definition of Axiom A no-cycle diffeomorphisms coincide with mine. He only requires that there be no m-cycles, for $m \geq 2$.

Exercise 9.2 comes from [1.16].

It is sometimes easier to check for transversality than to show that there are no cycles.

References

[9.1] Palis, J., A note on Ω-stability, in *Global Analysis*, Vol. XIV (Proceedings of Synposia in Pure Mathematics), American Mathematical Society, Providence, R. I., 1970, p. 221.

[9.2] Robinson, C., C^r structural stability implies Kupka–Smale, in *Dynamical Systems*, Peixoto (Ed.), Academic Press, New York, 1973, p. 443.

[9.3] Smale, S., Stability and genericity in dynamical systems, *Seminaire Bourbaki*, 1969—1970, Lecture Notes in Mathematics, No. 180, Springer-Verlag, New York, 1971.

CHAPTER 10

Markov Partitions

I. Symbolic Dynamics

Our last major result will be counting the periodic points in a hyperbolic set with local product structure; we will carry this out using the important technique of symbolic dynamics.

Let k be an integer greater than zero. We denote by $[k]$ the set of symbols $\{1,\ldots,k\}$ with the discrete topology; we denote by $\Sigma(k)$ the product space $[k]^{\mathbb{Z}}$. An element of $\Sigma(k)$ is an infinite sequence of symbols in $[k]$, $\underline{a} = \ldots a_{-1} a_0 a_1 \ldots$.

Note that the product topology on $\Sigma(k)$ is also induced by the metric $d(\underline{a}, \underline{b}) = \sum_{n=-\infty}^{\infty} 2^{-(2|n|+1)} \delta_n(\underline{a}, \underline{b})$, where $\delta_n(\underline{a}, \underline{b})$ is 0 when $a_n = b_n$, and 1 otherwise. From this we see that a sequence $(\underline{a}^j)_{j\in\mathbb{N}}$ converges if and only if, for every i, the sequence $(a_i^j)_{j\in\mathbb{N}}$ converges. Notice also that we have $d(\underline{a}, \underline{b}) \geq \frac{1}{2} \Leftrightarrow a_0 \neq b_0$. $\Sigma(k)$ is compact, totally disconnected and has no isolated points, hence is a Cantor set.

We can define a homeomorphism of $\Sigma(k)$, the shift σ, by stipulating that the shift moves the sequence one place to the left, $(\sigma(\underline{a}))_i = a_{i+1}$. We are now ready to define certain zero-dimensional analogues of hyperbolic sets. Denote by M_k the set of all $k \times k$ matrices all of whose entries are 0 or 1.

Definition 10.1. Let $A = (A_{ij}) \in M_k$. We then define

$$\Sigma_A = \{\underline{a} \in \Sigma(k) | \forall i\, A_{a_i a_{i+1}} = 1\}$$

Σ_A is a closed σ-invariant subspace of $\Sigma(k)$; we denote by σ_A and often just σ the restriction of σ to Σ_A. The pair (Σ_A, σ_A) is called a *subshift of finite type.*

Example. Encoding a map with a partition.

Let $P_1, \ldots, P_k$ be a partition of a set E, and let f be a map from E to E. We define a matrix $A = (A_{ij}) \in M_k$ by

$$A_{ij} = 1 \Leftrightarrow f(P_i) \cap (P_j) \neq \emptyset.$$

To each element x of E we can associate a string $\underline{a} = \tau(x)$ in Σ_A by following x's itinerary among the P's:

$$\forall i \in \mathbb{Z}, \quad P_{a_i} \quad \text{contains} \quad f^i(x).$$

With this definition, the following diagram commutes:

$$\begin{array}{ccc} E & \xrightarrow{\tau} & \Sigma_A \\ {\scriptstyle f}\downarrow & & \downarrow{\scriptstyle \sigma} \\ E & \xrightarrow{\tau} & \Sigma_A. \end{array}$$

Given $A \in M_k$, we say a finite string $a_0 \ldots a_n$ of symbols in $[k]$ is admissible for A if and only if $A_{a_i a_{i+1}} = 1$, $i = 0, \ldots, n - 1$. For any two symbols p and q in $[k]$ we denote by $N_n(p, q, A)$ the number of admissible strings for A of length $n + 1$, beginning with p and ending with q. We can readily compute this number.

Lemma 10.2. $N_n(p, q, A) = (A^n)_{p,q}$.

Proof. For $n = 1$, this is clear from the definition. Suppose, then, that it is true for $n = m - 1$. The next to last term, r say, of an admissible sequence can only be a symbol with $A_{rq} = 1$. Consequently,

$$\begin{aligned} N_m(p, q, A) &= \sum_{r=1}^{k} N_{m-1}(p, r, A) A_{rq} \\ &= \sum_{r=1}^{k} (A^{m-1})_{p,r} A_{rq} = (A^m)_{p,q}, \end{aligned}$$

and we are done by induction. □

If f is a continuous map of a topological space to itself we denote by $N_n(f)$ the number of isolated fixed points of f^n. Lemma 10.2 allows us, as a corollary, to count the periodic points of a subshift of finite type.

Proposition 10.3. $N_n(\sigma_A) = \operatorname{tr} A^n$.

Subshifts of finite type also have a sort of local product structure. First we will define the local stable and unstable sets:

$$W^s_{1/3}(\underline{a}) = \{\underline{b} \in \Sigma_A \mid \forall n \geq 0, d(\sigma^n(\underline{a}), \sigma^n(\underline{b})) \leq \tfrac{1}{3}\},$$

$$W^u_{1/3}(\underline{a}) = \{\underline{b} \in \Sigma_A \mid \forall n \geq 0, d(\sigma^n(\underline{a}), \sigma^n(\underline{b})) \leq \tfrac{1}{3}\}.$$

The following proposition is the result of simple calculations.

Proposition 10.4. *For all $\underline{a}$ in Σ_A,*

$$W^s_{1/3}(\underline{a}) = \{\underline{b} | \forall n \geq 0, a_n = b_n\},$$

$$W^u_{1/3}(\underline{a}) = \{\underline{b} | \forall n \geq 0, a_n = b_n\}.$$

If the distance from $\underline{a}$ to $\underline{b}$ is less than or equal to $\frac{1}{3}$, then $W^s_{1/3}(\underline{a}) \cap W^u_{1/3}(\underline{b})$ is exactly one point $\underline{c}$, for which $c_n = a_n$, $n \geq 0$ and $c_n = b_n$, $n \leq 0$. The shift σ_A is expansive, with constant of expansivity $\frac{1}{3}$. Moreover, σ contracts W^s and expands W^u

$$\forall \underline{b} \in W^s_{1/3}(\underline{a}), \qquad d(\sigma(\underline{a}), \sigma(\underline{b})) = \tfrac{1}{4} d(\underline{a}, \underline{b}),$$

$$\forall \underline{b} \in W^u_{1/3}(a), \qquad d(\sigma^{-1}(\underline{a}), \sigma^{-1}(\underline{b})) = \tfrac{1}{4} d(\underline{a}, \underline{b}).$$

Let $U_{1/2}$ be the open neighborhood of the diagonal in $\Sigma_A \times \Sigma_A$ defined by

$$U_{1/2} = \{(\underline{a}, \underline{b}) \in \Sigma_A \times \Sigma_A | d(\underline{a}, \underline{b}) < \tfrac{1}{2}\}.$$

We can, then, define a map, the bracket,

$$[\cdot, \cdot]: U_{1/2} \to \Sigma_A,$$

$$[\underline{a}, \underline{b}] = \underline{c} = W^u_{1/3}(\underline{a}) \cap W^s_{1/3}(\underline{b}).$$

Proposition 10.5. *Let $\underline{a}$ be a point of Σ_A. The bracket is a homeomorphism of $W^u_{1/3}(\underline{a}) \times W^s_{1/3}(\underline{a})$ onto the open set $U(\underline{a}) = \{\underline{b} | a_0 = b_0\}$.*

Proof. The continuity of $[\cdot, \cdot]$ is clear from Proposition 10.4, as is the fact that

$$[W^u_{1/3}(\underline{a}), W^s_{1/3}(\underline{a})] \subset U(\underline{a}).$$

Notice that if $\underline{b}$ is in $U(\underline{a})$ then $(\underline{a}, \underline{b})$ is in $U_{1/2}$, so it makes sense to define a map

$$U(\underline{a}) \to W^u_{1/3}(\underline{a}) \times W^s_{1/3}(\underline{a}),$$

$$\underline{b} \to ([\underline{b}, \underline{a}], [\underline{a}, \underline{b}]).$$

Since $[\cdot, \cdot]$ is continuous this map is as well, and it is clear that it is the inverse of the bracket. □

Next we will develop a handy way of keeping track of the sequence $N_n(f)$, the ζ-function.

Definition 10.6. Let $\alpha = (\alpha_n)_{n \geq 1}$ be a sequence of complex numbers. We denote by ζ_α the formal series

$$\zeta_\alpha(x) = \exp\left(\sum_{n=1}^{\infty} \frac{\alpha_n}{n} t^n\right).$$

If f is continuous map of a topological space to itself we set

$$\zeta_f(t) = \exp\left(\sum_{n=1}^{\infty} \frac{N_n(f)}{n} t^n\right).$$

The ζ-function has some very nice properties:

(1) $\zeta_{\alpha+\beta}(t) = \zeta_\alpha(t)\zeta_\beta(t)$; and
(2) if $\alpha = (\lambda^n)_{n\geq 1}$, then $\zeta_\alpha = 1/(1-\lambda t)$ since $\log(1/(1-\lambda t)) = \sum_{n=1}^{\infty} (\lambda^n/n)t^n$. This allows us to demonstrate the following simple proposition which is at the heart of the ζ-function's usefulness.

Proposition 10.7. *Let A and B be two square matrices. If* $\alpha = (\operatorname{tr} A^n - \operatorname{tr} B^n)_{n\geq 1}$, *then*

$$\zeta_\alpha(t) = \frac{\det(I - tB)}{\det(I - tA)}.$$

PROOF. By (1) it is enough to consider the case when $B = 0$. Suppose A is $k \times k$ and let $\lambda_1, \ldots, \lambda_k$ be its eigenvalues.

From (2) we have

$$\begin{aligned}\zeta_\alpha(t) &= \exp\left(\sum_{n=1}^{\infty} \frac{\operatorname{tr} A^n}{n} t^n\right)\\ &= \exp\left(\sum_{n=1}^{\infty}\left(\sum_{m=1}^{k} \lambda_m^n\right)\frac{t^n}{n}\right)\\ &= \prod_{m=1}^{k} \frac{1}{1-\lambda_m t} = \frac{1}{\det(I - tA)}.\end{aligned}$$ □

In particular, from Proposition 10.3, we have:

Corollary 10.8. $\zeta_{\sigma_A}(t) = 1/\det(I - tA)$.

Proposition 10.9. (1) *If A is a* $k \times k$ *complex matrix with eigenvalues* $\lambda_1, \ldots, \lambda_k$, *we have*

$$\limsup_{n\to+\infty} \frac{1}{n}\log|\operatorname{tr} A^n| = \max_i \log|\lambda_i|.$$

(2) *If, moreover, A has integral entries and the quantity in* (1) *is zero, then the* λ*'s are either zero or roots of unity.*

PROOF. (1) Set $\alpha = (\operatorname{tr} A^n)_{n\geq 1}$. Since $\zeta_\alpha(t) = 1/\det(I - tA)$, the radius of convergence $\rho(\zeta_\alpha)$ is $(\max_i |\lambda_i|)^{-1}$. Since the exponential is an entire function, the radius of convergence of $\sum_{n=1}^{\infty} (\operatorname{tr} A^n/n)t^n$ is no greater than $\rho(\zeta_\alpha)$. Therefore

$$\left(\limsup_{n\to+\infty} \sqrt[n]{\frac{\operatorname{tr} A^n}{n}}\right)^{-1} \leq \left(\max_i |\lambda_i|\right)^{-1}$$

so

$$\max_i \log|\lambda_i| \leq \limsup_{n\to+\infty} \frac{1}{n}\log|\operatorname{tr} A^n|.$$

On the other hand

$$\forall n \in \mathbb{N}, \qquad |\mathrm{tr}\, A^n| \le k\left(\max_{i \in [k]} |\lambda_i|\right)^n$$

so we must have the desired equality.

(2) Since $\max_i \log|\lambda_i| = 0$, the numbers $\lambda_1, \ldots, \lambda_k$ all lie in the unit disc and are roots of a monic integral polynomial of degree k, and the same is true for the numbers $\lambda_1^n, \ldots, \lambda_k^n$, for every positive integer n. There are only a finite number of such polynomials, since the requirement that the roots lie in the unit disc imposes bounds on the coefficients of a monic integral polynomial. Therefore, there are only a finite number of possible roots, so for each i, the set $\{\lambda_i^n | n \ge 1\}$ is finite and hence λ_i is either 0 or a root of unity. □

II. Markov Partitions

We now proceed to define and construct Markov partitions for closed hyperbolic invariant sets with local product structure. First we fix our notations and assumptions. Λ will be a closed hyperbolic set for a C^r diffeomorphism ($r \ge 1$) of a manifold M. We will assume that Λ has a local product stucture

$$\exists \varepsilon, \delta > 0 \qquad \text{such that} \quad \forall x, y \in \Lambda,$$

$$d(x, y) < \delta \Rightarrow [x, y] = W_\varepsilon^s(x) \cap W_\varepsilon^u(y) \in \Lambda,$$

where $\delta < \varepsilon/2$ and ε is a constant of expansivity for f on Λ. If A is a subset of Λ, we will denote by $\mathring{A}$ its interior as a subspace of Λ; ∂A will be its frontier in Λ. When x is in Λ, we define

$$W^s(x, A) = W^s(x) \cap A.$$

When η is a constant less than or equal to ε we set

$$W_\eta^s(x) = \{y \in W_\varepsilon^s(x) | d(x, y) \le \eta\}$$

and

$$\mathring{W}_\eta^s(x) = \{y \in W_\varepsilon^s(x) | d(x, y) < \eta\}.$$

If B is a subset of $W_\varepsilon^s(x) \cap \Lambda$ we denote by $\mathrm{Int}(B)$ its interior as a subset of $W_\varepsilon^s(x) \cap \Lambda$ and by $\mathrm{fr}(B)$ its frontier. We have analogous definitions and notations for the unstable manifolds. Notice that with ε and δ as above, it is immediate from the stable manifold theorem that the restriction of f to $\mathring{W}_\delta^s(x)$ is an open mapping to $W_\varepsilon^s(f(x))$, for any x in Λ, and similarly for f^{-1} and W^u.

We have the following parallel to Proposition 10.5.

Proposition 10.10. *There is a positive constant ρ, less than $\delta/2$, such that for all x in Λ, the bracket is a homeomorphism of the product $(\mathring{W}_\rho^u(x) \cap \Lambda) \times (\mathring{W}_\rho^s(x) \cap \Lambda)$ onto an open neighborhood of x in Λ.*

PROOF. Let $V_\delta(x) = \{y \in \Lambda \mid d(x, y) < \delta\}$. We define two continuous maps on V_δ.

$$\Pi_s\colon V_\delta(x) \to W^s_\varepsilon(x, \Lambda),$$
$$y \mapsto [x, y],$$
$$\Pi_u\colon V_\delta(x) \to W^u_\varepsilon(x, \Lambda),$$
$$y \mapsto [y, x].$$

Now the bracket is uniformly continuous on a compact neighborhood of the diagonal in $\Lambda \times \Lambda$ so

$$\exists \rho > 0, \qquad \rho < \delta/2$$

such that

$$\forall x, y, z \in \Lambda,$$
$$d(x, y) < \rho,$$

and

$$d(x, z) < \rho \Rightarrow d(x, [y, z]) < \delta.$$

Thus if $(y, z) \in (\mathring{W}^u_\rho(x) \cap \Lambda) \times (\mathring{W}^s_\rho(x) \cap \Lambda)$, Π_s and Π_u are defined and (Π_u, Π_s) is an inverse for $[\cdot, \cdot]$. The bracket is therefore a homeomorphism on the open set $\Pi_s^{-1}(\mathring{W}^s_\rho(x) \cap \Lambda) \cap \Pi_u^{-1}(\mathring{W}^u_\rho(x) \cap \Lambda)$. □

Definition 10.11. A subset R of Λ is called a *rectangle* if it has a diameter less than δ and is closed under the bracket, i.e.,

$$x \in R, \qquad y \in R \Rightarrow [x, y] \in R.$$

A rectangle R is *proper* if it is the closure of its interior in Λ; $\bar{\mathring{R}} = R$.

Using Proposition 10.10, the following is clear.

Proposition 10.12. *If R is a rectangle of diameter less than ρ and x is a point of R, then*

$$\mathring{R} = [\operatorname{Int} W^u(x, R), \operatorname{Int} W^s(x, R)],$$
$$\partial R = [\operatorname{fr} W^u(x, R), W^s(x, R)] \cup [W^u(x, R), \operatorname{fr} W^s(x, R)].$$

Corollary 10.13. *Let R be a rectangle of diameter less than ρ. Then*:

(1) $x \in \mathring{R}$ *if and only if* $x \in \operatorname{Int}(W^u(x, R)) \cap \operatorname{Int}(W^s(x, R))$;
(2) *for* $x \in \mathring{R}$, $\operatorname{Int} W^s(x, R) = W^s(x, \mathring{R})$ *and* $\operatorname{Int} W^u(x, R) = W^u(x, \mathring{R})$;
(3) *if R is closed then $\partial R = \partial^s R \cup \partial^u R$, where*

$$\partial^s R = \{x \in R \mid x \notin \operatorname{Int} W^u(x, R)\} = \{x \in R \mid W^s(x, R) \cap \mathring{R} = \varnothing\},$$
$$\partial^u R = \{x \in R \mid x \notin \operatorname{Int} W^s(x, R)\} = \{x \in R \mid W^u(x, R) \cap \mathring{R} = \varnothing\}.$$

PROOF. Conclusions (1) and (2) are immediate. For (3), notice that for any z in R,

$$[z, y] = y \quad \text{whenever} \quad y \in W^s(z, R),$$
$$[y, z] = y \quad \text{whenever} \quad y \in W^u(z, R).$$

Therefore

$$[\{z\} \cap \text{Int } W^u(z, R), \text{Int } W^s(z, R)] = \mathring{R} \cap \text{Int } W^s(z, R) = \mathring{R} \cap W^s(z, R),$$
$$[\text{Int } W^u(z, R), \{z\} \cap W^u(z, R)] = \text{Int } W^u(z, R) \cap \mathring{R} = W^u(z, R) \cap \mathring{R},$$

and (3) follows. □

In order to apply Proposition 10.12 and Corollary 10.13 we henceforth assume without comment that all *small* rectangles have diameter less than ρ.

Proposition 10.14. *Let R be a small closed rectangle and δ' a constant between 0 and δ. Then the set $\{x \in \Lambda \mid W^s_{\delta'}(x) \cap \partial^s R = \varnothing\}$ is open and dense in Λ. Moreover, if $W^s_{\delta'}(x) \cap \partial^s R$ is empty, then $(W^s_{\delta'}(x) \cap \Lambda) - \partial R$ is open and dense in $W^s_{\delta'}(x) \cap \Lambda$. Analogous results hold for the unstable manifolds.*

PROOF. The set $W^s(x, R)$ depends continuously on x in R; consequently $\partial^s R = \{x \in R \mid W^s(x, R) \cap \mathring{R} = \varnothing\}$ is closed. Since $W^s_{\delta'}(x)$ also depends continuously on x in Λ, the set $\{x \in \Lambda \mid W^s_{\delta'}(x) \cap \partial^s R = \varnothing\}$ is open in Λ.

We now show that this last is also dense in Λ. Let y be in $\partial^s R \cap W^s_{\delta'}(x)$. By Corollary 10.13, y belongs to the frontier of $W^u(y, R)$ in $W^u_\varepsilon(y) \cap \Lambda$. We have $[y, x] = x$ and $d(x, y) < \delta'$, so by the continuity of the bracket we can find a point z in $W^u_\varepsilon(y) \cap (\Lambda - R)$ close enough to y that

$$d(x, z) < \delta \quad \text{and} \quad d([z, x]), y) < \delta.$$

We then have

$$t \in W^s_{\delta'}([z, x]) \cap R \Rightarrow z = [[z, x], y] = [t, y] \in R.$$

Therefore $W^s_{\delta'}([z, x]) \cap R$ is empty. Since $[z, x]$ is arbitrarily close to x when z is arbitrarily close to y, we have demonstrated the density of $\{x \in \Lambda \mid W^s_{\delta'} \cap \partial^s R = \varnothing\}$.

Now we attack the second part. Let x be a point of Λ with $W^s_{\delta'}(x) \cap \partial^s R = \varnothing$. If $\mathring{W}^s_{\delta'}(x) \cap \partial^u R$ is also empty, then $\mathring{W}^s_{\delta'}(x) \cap \Lambda - \partial R$ is clearly dense in $\mathring{W}^s_{\delta'}(x) \cap \Lambda$. If not, assume that y belongs to $\mathring{W}^s_{\delta'}(x) \cap \Lambda \cap (\partial^u R)$. Corollary 10.13 implies that y belongs to fr $W^s(y, R))$, but $\mathring{W}^s_{\delta'}(x)$ is open in $W^s_\varepsilon(x)$ so there are points z, arbitrarily close to y, such that

$$z \in \mathring{W}^s_{\delta'}(x) \cap W^s_\varepsilon(y) \cap \Lambda - W^s(y, R).$$

These points cannot belong to R. Since ∂R is closed, this shows that $\mathring{W}^s_{\delta'}(x) \cap \Lambda - \partial R$ is open and dense in $\mathring{W}^s_{\delta'}(x) \cap \Lambda$. □

Definition 10.15. *A Markov Partition of* Λ for f is a finite collection $\mathbf{R} = \{R_1, \ldots, R_n\}$ of small proper rectangles with disjoint interiors with the property that if $f(R_i)$ or $f^{-1}(R_i)$ hit R_j, then they extend all the way across it. More precisely, we require that:

(i) $x \in \mathring{R}_i, \quad f(x) \in \mathring{R}_j \Rightarrow f(W^s(x, R_i)) \subset W^s(f(x), R_j);$

(ii) $x \in \mathring{R}_i, \quad f^{-1}(x) \in \mathring{R}_j \Rightarrow f^{-1}(W^u(x, R_i)) \subset W^u(f^{-1}(x), R_j).$

REMARKS. (1) The R's are implicitly assumed to have diameter less than ρ; and, furthermore, being proper, they are closed.

(2) By Corollary 10.13 and the above-mentioned openness of f on small stable discs, properties (i) and (ii) above are equivalent to:

(i′) $x \in \mathring{R}_i, \quad f(x) \in \mathring{R}_j \Rightarrow f(W^s(x, \mathring{R}_i)) \subset W^s(f(x), \mathring{R}_j);$

(ii′) $x \in \mathring{R}_i, \quad f(x) \in \mathring{R}_j \Rightarrow f^{-1}(W^u(x, \mathring{R}_i)) \subset W^u(f^{-1}(x), \mathring{R}_j).$

Example 10.16. Let A be a $k \times k$ matrix of 0's and 1's and (Σ_A, σ_A) the associated subshift of finite type. Let $C_i = \{\underline{a} \in \Sigma_A | a_0 = i\}$. Proposition 10.4 says, in effect, that $\mathbf{C} = \{C_1, \ldots, C_k\}$ is a Markov partition of Σ_A for σ_A. In a sense that will become clear later, this example is a universal model of all Markov partitions.

Example 10.17. Recall the Anosov diffeomorphism f of $T^2 = \mathbb{R}^2/\mathbb{Z}^2$ induced by the matrix $M = \begin{pmatrix} 2 & 1 \\ 1 & 1 \end{pmatrix}$ which was introduced in Example 4.6. The eigenvalues of M are

$$\lambda_s = \tfrac{1}{2}(3 - \sqrt{5}) \quad \text{and} \quad \lambda_u = \tfrac{1}{2}(3 + \sqrt{5}).$$

The corresponding stable and unstable manifolds, lifted to $\mathbb{R}^2$, are

$$y - y_0 = -\tfrac{1}{2}(1 + \sqrt{5})(x - x_0) \qquad \text{for} \quad W^s(x_0, y_0),$$

$$y - y_0 = -\tfrac{1}{2}(\sqrt{5} - 1)(x - x_0) \qquad \text{for} \quad W^u(x_0, y_0).$$

The torus T^2 can be decomposed into two closed sets R_1 and R_2 with disjoint interiors, which are covered by two parallelograms in R_2 with sides parallel to the eigendirections of M; see Figure 10.1. (T^2, f) has a local product structure and we could calculate the constants ε, δ, and ρ for this structure.

It is an easy, but tedious, exercise then to check that by taking N large enough, the collection of rectangles $\mathbf{R} = \{\bigcap_{i=-N}^{N} f^i(R_{\varepsilon(i)}), \varepsilon(i) \in \{1, 2\}\}$ each have diameter less than ρ and in fact are Markov partitions of T^2 for f.

Let us show that there is always a Markov partition for any hyperbolic set Λ with local product structure. First we supplement the notation we have already.

Let β be a constant $0 < \beta < \rho/2 < \delta/4 < \varepsilon/8$. Using Proposition 8.21 we

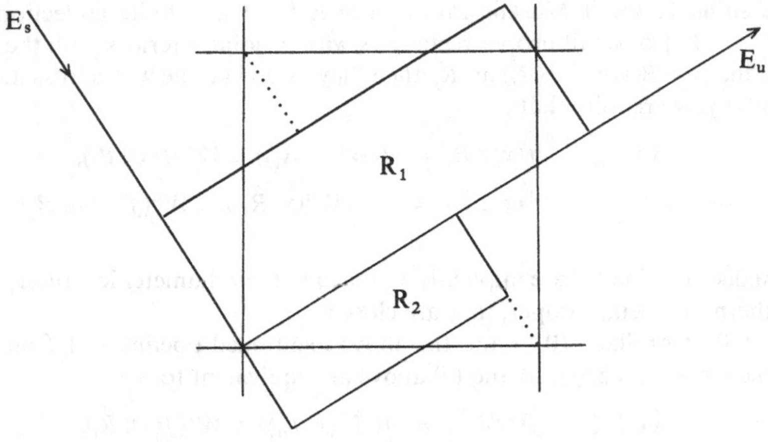

Figure 10.1.

can find a constant α such that $0 < \alpha < \beta$ and every α-psuedo-orbit in Λ is β-shadowed by a unique orbit in Λ. Furthermore, since Λ is compact, we can find a constant γ, $0 < \gamma < \alpha/2$, such that for all $x, y \in \Lambda$ with $d(x, y) < \gamma$ we have $d(f(x), f(y)) < \alpha/2$, and we can cover Λ with a finite number of open balls of radius less than γ and centers $p_1, \ldots, p_k$, say. We define a matrix $A \in M_k$ by

$$A_{ij} = 1 \quad \text{if} \quad f(p_i) \in \mathring{B}_\alpha(p_j), \qquad A_{ij} = 0 \quad \text{otherwise.}$$

For every $\underline{a}$ in Σ_A, the sequence $(p_{a_i})_{i \in \mathbb{Z}}$ is an α-psuedo-orbit; it is therefore β-shadowed by a unique point $x = \theta(\underline{a})$. This defines a map $\theta\colon \Sigma_A \to \Lambda$.

Proposition 10.18. *The mapping θ is a semiconjugacy of σ_A and f; i.e., it is continuous, surjective, and makes the following diagram commute:*

$$\begin{array}{ccc} \Sigma_A & \xrightarrow{\sigma_A} & \Sigma_A \\ \theta \downarrow & & \downarrow \theta \\ \Lambda & \xrightarrow{f} & \Lambda. \end{array}$$

PROOF. If x β-shadows (p_{a_i}), then $f(x)$ β-shadows $(p_{(\sigma\underline{a})_i})$, so the diagram commutes.

If θ were not continuous, the compactness of Λ would allow us to find two sequences $(\underline{s}^n)$ and $(\underline{t}^n)$ with the same limit l in Σ_A whose images under θ had distinct limits s and t in Λ. However, we know that

$$\forall i \in \mathbb{Z}, \qquad \forall n \in \mathbb{N}, \qquad d(f^i(\theta(\underline{s}^n)), p_{s_i^n}) < \beta,$$

so passing to the limit we have

$$\forall i \in \mathbb{Z}, \qquad d(f^i(s), p_{l_i}) \le \beta, \qquad d(f^i(t), p_{l_i}) \le \beta.$$

Therefore,

$$\forall i \in \mathbb{Z}, \qquad d(f^i(s), f^i(t)) \le 2\beta < \delta/2 < \varepsilon,$$

but ε is a constant of expansivity so we reach the contradiction that $s = t$.

To show, finally, that θ is surjective, let x be a point of Λ and $\underline{a}$ an element of Σ_A such that

$$\forall i \in \mathbb{Z}, \qquad f^i(x) \in \mathring{B}_\gamma(p_{a_i}).$$

Then, by the definition of γ, we have

$$d(f(p_{a_i}), p_{a_{i+1}}) \le d(f(p_{a_i}), f(f^i(x))) + d(f^{i+1}(x), p_{a_{i+1}}) \le \alpha/2 + \gamma < \alpha.$$

In other words, $\underline{a}$ belongs to Σ_A; since $\theta(\underline{a}) = x$, we are done. □

Proposition 10.19. *The map θ is a morphism of the local product structure. More precisely, we have*

$$\forall \underline{a} \in \Sigma_A, \qquad \theta(W^s_{1/3}(\underline{a})) \subset W^s_\varepsilon(\theta(\underline{a})),$$
$$\theta(W^u_{1/3}(\underline{a})) \subset W^u_\varepsilon(\theta(\underline{a})),$$
$$d(\underline{a}, \underline{b}) < \tfrac{1}{2} \Rightarrow d(\theta(\underline{a}), \theta(\underline{b})) < \rho \le \delta,$$

and

$$\theta([\underline{a}, \underline{b}]) = [\theta(\underline{a}), \theta(\underline{b})].$$

Proof. Recall that whenever $d(\underline{a}, \underline{b}) < \frac{1}{2}$, we have $a_0 = b_0$. Hence

$$d(\theta(\underline{a}), \theta(\underline{b})) \le d(\theta(\underline{a}), p_{a_0}) + d(p_{a_0}, \theta(\underline{b})) \le 2\beta < \rho,$$

so $[\theta(\underline{a}), \theta(\underline{b})]$ is defined.

Let $\underline{a}$ be an element of Σ_A, and $\underline{c}$ a point of $W^u_{1/3}(\underline{a})$. Proposition 10.4 then tells us that $c_i = a_i$, for all $i \ge 0$. Consequently, we have

$$\forall i \ge 0, \qquad d(f^i(\theta(\underline{a})), f^i(\theta(\underline{c}))) \le d(f^i(\theta(\underline{a})), p_{a_i}) + d(p_{c_i}, f^i(\theta(\underline{c}))) \le 2\beta < \varepsilon,$$

in other words, $\theta(\underline{c}) \in W^s_\varepsilon(\theta(\underline{a}))$.

Similarly, we can show that $\theta(W^u_{1/3}(\underline{a})) \subset W^u_\varepsilon(\theta(\underline{a}))$.

Combining this with the definitions of the brackets, we obtain

$$\theta([\underline{a}, \underline{b}]) = \theta(W^s_{1/3}(\underline{a}) \cap W^u_{1/3}(\underline{b})) \subset W^s_\varepsilon(\theta(\underline{a})) \cap W^u_\varepsilon(\theta(\underline{b})) = [\theta(a), \theta(\underline{b})].$$

□

Setting $T_i = \theta(C_i)$, where C_i are as in Example 10.16, we see that the T_i's cover Λ, since θ is surjective.

Lemma 10.20. *Let $i \in [k]$ and $\underline{a}$ a point of C_i. Then T_i is a closed rectangle of diameter less than ρ and we have*

$$\theta(W^s(\underline{a}, C_i)) = W^s(\theta(\underline{a}), T_i),$$
$$\theta(W^u(\underline{a}, C_i)) = W^u(\theta(\underline{a}), T_i).$$

Proof. Since C_i is a rectangle of diameter less than $\frac{1}{2}$, the previous lemma implies that T_i is a rectangle of diameter less than ρ; T_i is closed since C_i is compact and θ is continuous.

Moreover, from the definition of C_i, we see that $W^s(\underline{a}, C_i) = W^s_{1/3}(\underline{a})$ so, by 10.19

$$\theta(W^s(\underline{a}, C_i)) \subset W^s_\varepsilon(\theta(\underline{a})) \cap T_i = W^s(\theta(\underline{a}), T_i)$$

and, similarly, $\theta(W^u(\underline{a}, C_i))$ must be contained in $W^u(\underline{a}), T_i)$.

Conversely, if x belongs to $W^s(\theta(\underline{a}), T_i)$, let $\underline{s}$ be a preimage of x under θ in C_i. We have

$$\theta([\underline{a}, \underline{s}]) = [\theta(\underline{a}), \theta(\underline{s})] = [\theta(\underline{a}), x] = x$$

and, furthermore, $[\underline{a}, \underline{s}]$ belongs to $W^s(\underline{a}, C_i)$. Therefore,

$$W^s(\theta(\underline{a}), T_i) \subset \theta(W^s(\underline{a}, C_i)),$$

and, likewise

$$W^u(\theta(\underline{a}), T_i) \subset \theta(W^u(\underline{a}, C_i)).$$ □

Proposition 10.21. *If $\underline{s}$ belongs to C_i and $\sigma(\underline{s})$ belongs to C_j, then setting $x = \theta(\underline{s})$, we have*

$$f[W^s(x, T_i)] \subset W^s(f(x), T_j),$$
$$f^{-1}[W^u(f(x), T_j)] \subset W^u(x, T_i).$$

Proof. Using Lemma 10.20, we have

$$\begin{aligned} f(W^s(x, T_i)) &= f(W^s(\theta(\underline{s}), T_i)) \\ &= f\theta(W^s(\underline{s}, C_i)) \\ &= \theta\sigma(W^s(\underline{s}, C_i)) \\ &\subset \theta(W^s(\sigma(\underline{s}), C_j)) \\ &= W^s(\theta(\sigma(s)), T_j) \\ &= W^s(f(x), T_j). \end{aligned}$$

The proof for W^u is analogous. □

Corollary 10.22. *Let x be a point of T_i. Then:*

(1) $\exists j \in [k]$ *such that*

$$f(x) \in T_j,$$
$$f(W^s(x, T_i)) \subset W^s(f(x), T_j),$$
$$f^{-1}(W^u(f(x), T_j)) \subset W^u(x, T_i).$$

(2) $\exists m \in [k]$ *such that*

$$f^{-1}(x) \in T_m,$$
$$f(W^s(f^{-1}(x), T_m)) \subset W^s(x, T_i),$$
$$f^{-1}(W^u(x, T_i)) \subset W^u(f^{-1}(x), T_m).$$

Proof. Take $\underline{s} \in C_i$ such that $\theta(\underline{s}) = x$. We define j and m by $\sigma(\underline{s}) \in C_j$ and $\sigma^{-1}(\underline{s}) \in C_m$, and apply the preceding proposition. □

The Markov partition we seek is obtained by subdividing the T_i's. When T_i and T_j intersect, we set

$$\tau_{ij}^1 = \{x \in T_i \mid W^s(x, T_i) \cap T_j \neq \emptyset,\ W^u(x, T_i) \cap T_j \neq \emptyset\},$$
$$\tau_{ij}^2 = \{x \in T_i \mid W^s(x, T_i) \cap T_j \neq \emptyset,\ W^u(x, T_i) \cap T_j = \emptyset\},$$
$$\tau_{ij}^3 = \{x \in T_i \mid W^s(x, T_i) \cap T_j = \emptyset,\ W^u(x, T_i) \cap T_j \neq \emptyset\},$$
$$\tau_{ij}^4 = \{x \in T_i \mid W^s(x, T_i) \cap T_j = \emptyset,\ W^u(x, T_i) \cap T_j = \emptyset\},$$

Lemma 10.23. *The* $\tau_{ij}^m (m = 1, 2, 3, 4)$ *form a partition of* T_i. *Moreover,* $\tau_{ij}^1 = T_i \cap T_j$ *and the sets* τ_{ij}^1, $\tau_{ij}^1 \cup \tau_{ij}^2$, *and* $\tau_{ij}^1 \cup \tau_{ij}^3$ *are all closed.*

Proof. The first assertion is obvious, as is the fact that $T_i \cap T_j \subset \tau_{ij}^1$. One the other hand, if $x \in \tau_{ij}^1$, choose y in $W^s(x, T_i) \cap T_j$ and z in $W^u(x, T_i) \cap T_j$; then $x = [y, z]$ and we see $x \in T_j \cap T_i$.

Since T_i and T_j are closed, so is their intersection, τ_{ij}^1. Moreover, we have

$$\tau_{ij}^1 \cup \tau_{ij}^2 = \{x \in T_i \mid W^s(x, T_i) \cap T_j \neq \emptyset\} = \{x \in T_i \mid W_\varepsilon^s(x) \cap (T_i \cap T_j) \neq \emptyset\},$$
$$\tau_{ij}^1 \cup \tau_{ij}^3 = \{x \in T_i \mid W^u(x, T_i) \cap T_j \neq \emptyset\} = \{x \in T_i \mid W_\varepsilon^u(x) \cap (T_i \cap T_j) \neq \emptyset\}.$$

Since $W_\varepsilon^s(x)$ and $W_\varepsilon^u(x)$ depend continuously on x and $T_i \cap T_j$ is closed, it follows that $\tau_{ij}^1 \cup \tau_{ij}^2$ and $\tau_{ij}^1 \cup \tau_{ij}^3$ are also closed. □

Now let $T_{ij}^m = \overline{\tau_{ij}^m}$, $m = 1, 2, 3, 4$.

Lemma 10.24. *The rectangle* T_i *decomposes as* $\bigcup_{m=1}^4 T_{ij}^m$, *with the following properties*:

(1) $T_{ij}^1 = T_i \cap T_j$.
(2) $\mathring{T}_{ij}^l \cap T_{ij}^m = \emptyset$ *if* $l < m$ *or* $(l, m) = (3, 2)$; *in particular the* T_{ij}^m*'s have disjoint interiors.*
(3) $T_{ij}^l \subset \tau_{ij}^l \cup (\bigcup_{m=1}^4 \partial T_{ij}^m)$.
(4) *The* T_{ij}^m*'s and their interiors are small rectangles* (see Figure 10.2).

Proof. Property (1) is just a restatement of part of Lemma 10.23. We treat the various cases of (2) separately:

$$l = 1, \qquad \mathring{T}_{ij}^1 = \mathring{\tau}_{ij}^1 \subset \tau_{ij}^1.$$

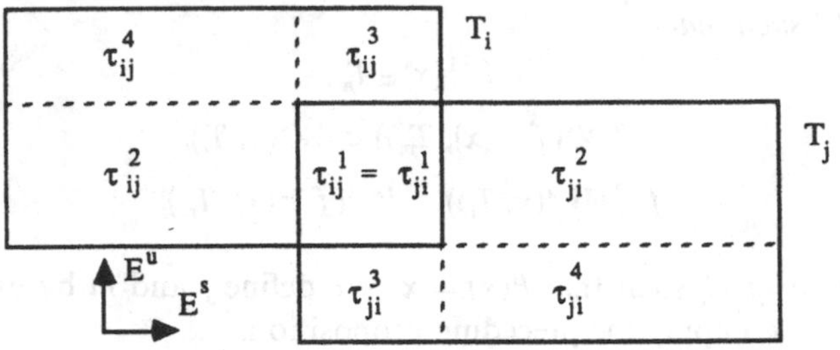

Figure 10.2.

Therefore, $\mathring{T}_{ij}^1 \cap \tau_{ij}^m = \emptyset$ whenever $m > 1$, so we see

$$\mathring{T}_{ij}^1 \cap T_{ij}^m = \mathring{T}_{ij}^1 \cap \overline{\tau_{ij}^m} = \emptyset, \qquad m > 1,$$

$$l = 2, \qquad \mathring{T}_{ij}^2 \subset T_{ij}^2 \subset \tau_{ij}^1 \cap \tau_{ij}^2.$$

Consequently, for $m > 2$: $\mathring{T}_{ij}^2 \cap T_{ij}^m = T_{ij}^2 \cap \overline{\tau_{ij}^m} = \emptyset$. The case $l = 3$ is treated similarly.

Since, for $l \neq m$, $\mathring{T}_{ij}^l \cap \mathring{T}_{ij}^m$ is empty, we have

$$\mathring{T}_{ij}^l - \left(\bigcup_{m=1}^{4} \partial T_{ij}^m \right) \subset T_i - \bigcup_{m \neq l} T_{ij}^m \subset \tau_{ij}^l$$

and (3) follows.

From Proposition 10.12 and the continuity of the bracket it follows that the closure of the interior of a rectangle of diameter less than ρ is also a rectangle of diameter less than ρ. To prove (4) it thus suffices to show that the τ_{ij}^m's are rectangles.

Now, if x and y belong to T_i, then

$$W^s([x, y], T_i) = W^s(x, T_i),$$

$$W^u([x, y], T_i) = W^u(y, T_i).$$

Therefore, when x and y belong to τ_{ij}^m, so does their bracket $[x, y]$. Since diam $\tau_{ij}^m <$ diam $T_i < \rho < \delta$, the τ_{ij}^m's and, hence, the T_{ij}^m's and $\mathring{T}_{ij}^m$'s as well, are small rectangles. □

Set $Z = \Lambda - \bigcup_{i,j=1}^{k} \bigcup_{m=1}^{4} \partial T_{ij}^m$; this is an open dense subset of Λ. The preceding lemma shows that

$$\forall ij \in [k], \qquad \forall m \in [4], \qquad Z \cap T_{ij}^m = Z \cap \mathring{T}_{ij}^m = Z \cap \tau_{ij}^m,$$

a fact which we will constantly use without comment in the rest of our argument.

When x belongs to Z, we set

$$K(x) = \{T_i | x \in T_i\},$$

$$K^*(x) = \{T_j | \exists T_i \in K(x) \text{ such that } T_i \cap T_j \neq \emptyset\},$$

$$R(x) = \bigcap \{\mathring{T}_{ij}^m | T_i \in K(x),\ T_j \in K^*(x),\ x \in T_{ij}^m\}.$$

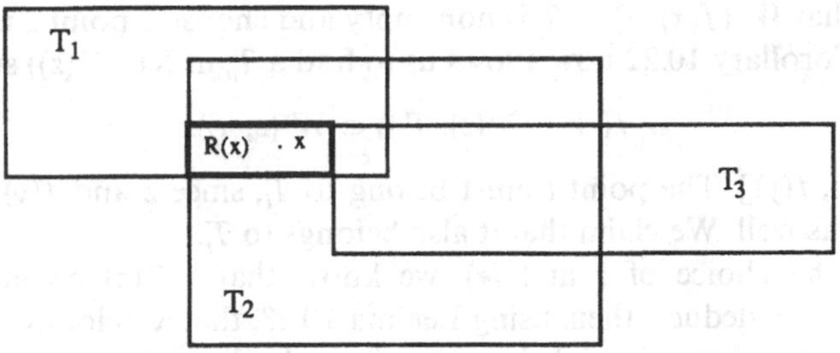

Figure 10.3.

$R(x)$ is clearly a small open rectangle which contains x as shown in Figure 10.3.

Lemma 10.25. *If x and y belong to Z the rectangles $R(x)$ and $R(y)$ are either identical or disjoint.*

PROOF. Let z belong to $Z \cap R(x)$. We will show that $R(z) = R(x)$, and for this, we will first show that $K(z) = K(x)$.

By the definition of $R(x)$ we see $K(x)$ is contained in $K(z)$. Conversely, if $T_j \in K(z)$ and $x \in T_i$, then T_j belongs to $K^*(x)$ and x belongs to some $\mathring{T}_{ij}^m$. Since z belongs to $\mathring{T}_{ij}^l$, by the preceding lemma, we must have $m = 1$ so that x also belongs to T_{ij} thus $T_j \in K(x)$. Therefore, $K(z) = K(x)$ and consequently $K^*(x) = K^*(z)$.

Now, for fixed i and j, the rectangles $\mathring{T}_{ij}^m$ are disjoint; therefore, we have $R(z) = R(x)$, whenever $z \in R(x) \cap Z$. If x and y belong to Z and $R(x) \cap R(y) \neq \varnothing$, then this intersection is open and hence contains a point z of Z. Therefore, we have $R(x) = R(z) = R(y)$. □

Lemma 10.26. *Let x and y be points of $Z \cap f^{-1}(Z)$ with $R(x) = R(y)$ and $y \in W_\varepsilon^s(x)$. Then we have $R(f(x)) = R(f(y))$.*

PROOF. We first show that $K(f(x)) = K(f(y))$.

If $f(x)$ belongs to T_i, then there is an $\underline{s}$ in C_i with $f(x) = \theta(\underline{s})$. Let j be s_{-1}. Since $x = \theta\sigma^{-1}(\underline{s})$, and $\theta(W^s(\sigma^{-1}(\underline{s}), C_j)) = W^s(x, T_j)$ by Lemma 10.20, the point y must have a preimage $\underline{t}$ under θ lying in $W^s(\sigma^{-1}(\underline{s}), C_j)$. Therefore, $t_1 = s_0 = i$. From this we deduce that $f(y) = \theta\sigma(y)$ belongs to T_i also. Thus we have $K(f(x)) \subset K(f(y))$. Since x and y play symmetric roles, this implies that $K(f(x)) = K(f(y))$ and hence $K^*(f(x)) = K^*(f(y))$.

Now fix a T_i in $K(f(x))$ and a T_j in $K^*(f(x))$. Since $W^s(f(x), T_i) = W^s(f(y), T_i)$, we have

$$W^s(f(x), T_i) \cap T_j \neq \varnothing \Rightarrow W^s(f(y), T_i) \cap T_j \neq \varnothing.$$

Let us now demonstrate a similar result for the unstable manifolds. By Corollary 10.22 we can find a rectangle T_m in $K(x)$ such that

$$(*) \qquad f^{-1}(W^u(f(x), T_i)) \subset W^u(x, T_m).$$

Assume that $W^u(f(x), T_i) \cap T_j$ is nonempty and choose a point z in this intersection. Corollary 10.22 now allows us to find a T_p in $K(f^{-1}(z))$ such that

$$(**) \qquad f(W^s(f^{-1}(z), T_p)) \subset W^s(z, T_j).$$

Let $t = [z, f(y)]$. The point t must belong to T_i, since z and $f(y)$ do, and to $W^u_\varepsilon(f(y))$ as well. We claim that it also belongs to T_j.

From the choice of z and (*), we know that $f^{-1}(z)$ belongs to $T_p \cap W^u(x, T_m)$. We deduce, then, using Lemma 10.23, that x belongs to $\tau^1_{mp} \cup \tau^3_{mp}$. Therefore, by assumption y belongs to $\tau^1_{mp} \cup \tau^3_{mp}$ also, which implies that

$$W^u(y, T_m) \cap T_p \neq \varnothing.$$

For any point u in this intersection we have

$$[f^{-1}(z), u] = [f^{-1}(z), y] \in W^s(f^{-1}(z), T_p).$$

So, (**) gives us

$$t = [z, f(y)] = f([f^{-1}(z), y]) \in W^s(z, T_j) \subset T_j,$$

and we have shown that

$$W^u(f(x), T_i) \cap T_j \neq \varnothing \;\Rightarrow\; W^u(f(y), T_i) \cap T_j \neq \varnothing.$$

Since x and y play equivalent roles in the proof the converse also holds. All together we have, then,

$$W^s(f(x), T_i) \cap T_j \neq \varnothing \;\Leftrightarrow\; W^s(f(y), T_i) \cap T_j \neq \varnothing,$$

$$W^u(f(x), T_i) \cap T_j \neq \varnothing \;\Leftrightarrow\; W^s(f(y), T_i) \cap T_j \neq \varnothing.$$

Since $f(x)$ and $f(y)$ are both in Z it follows that $R(f(x)) = R(f(y))$. □

The set of distinct $R(x)$'s for x in Z is finite since there are only finitely many $\mathring{T}^m_{ij}$'s. Let $\mathbf{R} = \{R_1, \ldots, R_s\} = \{\overline{R(x)} \mid x \in Z\}$.

Proposition 10.27. $\mathbf{R}$ *is a Markov partition of* Λ *for* f.

Proof. The R_i's cover Λ since they are closed and contain the dense set Z. They are proper since each is the closure of some open set $R(x)$, and one always has: the closure of the interior of the closure of the interior of A equals the closure of the interior of A. To see that R_i's have disjoint interiors, consider two distinct rectangles $R(x)$ and $R(y)$. Since each is open and they are disjoint by Lemma 10.25, we have

$$\begin{aligned} R(x) \cap R(y) = \varnothing &\Rightarrow R(x) \cap \overline{R(y)} = \varnothing \\ &\Rightarrow R(x) \cap \mathring{\overline{R(y)}} = \varnothing \\ &\Rightarrow \overline{R(x)} \cap \mathring{\overline{R(y)}} = \varnothing \\ &\Rightarrow \mathring{\overline{R(x)}} \cap \mathring{\overline{R(y)}} = \varnothing; \end{aligned}$$

in other words, the R_i's have disjoint interiors.

Finally, to check the Markov conditions 10.15(i) and (ii), we introduce the set Z^*

$$Z^* = \left\{x \in Z \mid W^s_{\delta/2}(x) \cap \left(\bigcup_{i,j=1}^{k} \bigcup_{m=1}^{4} \partial^s T^m_{ij}\right) = \varnothing\right\}.$$

Proposition 10.14 implies that Z^* is open and dense in Λ and that, for $x \in Z^*$, the set $\mathring{W}^s_{\delta/2}(x) \cap Z$ is open and dense in $\mathring{W}^s_{\delta/2} \cap \Lambda$. Since f is a homeomorphism on Λ, the set $Z^* \cap f^{-1}(Z^*)$ is open and dense in Λ.

For $x \in Z^* \cap f^{-1}(Z^*)$, the set $Z \cap \mathring{W}^s_{\delta/2}(x)$ is open and dense in $\mathring{W}^s_{\delta/2}(f(x)) \cap \Lambda$; therefore, $f^{-1}(z) \cap \mathring{W}^s_{\delta/2}(x)$ is dense in $\mathring{W}^s_{\delta/2}(x) \cap \Lambda$. Consequently, $Z \cap f^{-1}(Z) \cap \mathring{W}^s_{\delta/2}(x)$ is open and dense in $\mathring{W}^s_{\delta/2}(x) \cap \Lambda$.

Now suppose that x belongs to $Z^* \cap f^{-1}(Z^*) \cap \mathring{R}_i \cap f^{-1}(\mathring{R}_j)$; thus $R_i = \overline{R(x)}$ and $R_j = \overline{R(f(x))}$. The above reasoning tells us, in light of Proposition 10.10, that

$$W^s(x, R_i) = \overline{W^s(x, R(x))} = \overline{W^s(x, R(x)) \cap Z \cap f^{-1}(Z)}.$$

Now Lemma 10.26 implies that

$$f(W^s(x, R(x)) \cap Z \cap f^{-1}(Z)) \subset R(f(x)) \subset R_j;$$

by the continuity f, we then have $f(W^s(x, R_i)) \subset R_j$, so all together we see that

$$f(W^s(x, R_i)) \subset R_j \cap W^s_\varepsilon(f(x)) = W^s(f(x), R_j).$$

If x is an arbitrary point of $\mathring{R}_i \cap f^{-1}(\mathring{R}_j)$, then, since $Z^* \cap f^{-1}(Z^*)$ is dense and $\mathring{R}_i \cap f^{-1}(\mathring{R}_j)$ is open, there is a point x^* in $Z^* \cap f^{-1}(Z^*) \cap \mathring{R}_i \cap f^1(\mathring{R}_j)$. Since R_i is a rectangle, $W^s(x, R_i) = \{[x, z] \mid z \in W^s(x^*, R_i)\}$. From this it follows that

$$f(W^s(x, R_i)) \subset \{[f(x), f(z)] \mid z \in W^s(x^*, R_j)\}.$$

Applying the argument above to x^* shows that

$$f(W^s(x^*, R_i)) \subset W^s(f(x^*), R_j);$$

and thus we see, using the closure of R_j under the bracket, that

$$f(W^s(x^*, R_i)) \subset \{[f(x), w] \mid w \in W^s(f(x^*), R_j)\} = W^s(f(x), R_j).$$

In other words, we have shown that (i) holds:

$$\forall x \in \mathring{R}_i \cap f^{-1}(\mathring{R}_j), \qquad f(W^s(x, R_i)) \subset W^s(f(x), R_j).$$

The proof of (ii) is completely analogous. □

In conclusion, we have proved our existence theorem.

Theorem 10.28. *Let f be a C^r diffeomorphism ($r \geq 1$) of a manifold M. Let Λ be a closed hyperbolic invariant set for f with local product structure. Then Λ has a Markov partition for f.*

III. Applications of Markov Partitions

Next we examine the coding associated to our Markov partition; first we extend our fixed notations. Let $\mathbf{R} = \{R_1, \ldots, R_k\}$ be a Markov partition of Λ for f by small rectangles ($\forall i \in [k]$, diam $R_i < \rho < \delta/2 < \varepsilon/4$). We define a matrix A in M_k by

$$A_{ij} = 1 \quad \text{if} \quad f(\mathring{R}_i) \cap \mathring{R}_j \neq \varnothing, \qquad A_{ij} = 0 \quad \text{otherwise.}$$

Our immediate goal will be to define and study a canonical semiconjuacy Π from (Σ_A, σ_A) to (Λ, f). We will then show that the ζ-function of f restricted to Λ, ζ_f, is rational by counting the periodic points of (Σ_A, σ_A) and estimating the redundancy in the encoding of (Λ, f) by (Σ_A, σ_A). This project is similar to our discussion of the horseshoe in Chapter 4.

Definition 10.29. Let R be a rectangle of Λ. A subset S of R is called an *(un-) stable strip of R* if, for each x in S, $W^s(x, R) \subset S$ (resp. $W^u(x, R) \subset S$).

Proposition 10.30. *Let S be a stable strip of a rectangle R. Then S is a rectangle. If S is not empty, S intersects every $W^u(x, R)$ for x in R. Let x belong to R and y to S, then we have $S = [W^u(x, R) \cap S, W^s(y, R)]$.*

Proof. Let x be in R and y in S; then, by definition

$$[y, x] \in W^s(y, R) \subset S.$$

Therefore, S is a rectangle and intersects $W^u(x, R)$.

Now let z belong to S; since we evidently have $z = [[z, x], [y, z]]$, we see that

$$S \subset [W^u(x, R) \cap S, W^s(y, R)].$$

The opposite inclusion follows from S being a rectangle. □

Corollary 10.31. *If $\mathring{R}_i \cap f^{-1}(\mathring{R}_j)$ is not empty and if S is a nonempty stable strip of R_j (resp. $\mathring{R}_j$), then $f^{-1}(S) \cap R_i$ (resp. $f^{-1}(S) \cap \mathring{R}_i$) is a nonempty stable strip of R_i (resp. $\mathring{R}_i$).*

Proof. Let x be a point of $\mathring{R}_i \cap f^{-1}(\mathring{R}_j)$. $\mathbf{R}$ is a Markov partition, so

$$f^{-1}(W^u(f(x), R_j)) \subset W^u(x, R_i).$$

By the preceding proposition $S \cap W^u(f(x), R_j)$ is not empty, so, from above, $f^{-1}(S) \cap R_i$ is also not empty.

Letting y be a point of $f^{-1}(S) \cap R$ we have

$$W^s(y, R_i) = \{[y, z] \mid z \in W^s(x, R_i)\}.$$

Since $\mathbf{R}$ is a Markov partition $f(W^s(x, R_i)) \subset W^s(f(x), R_j)$; thus we see

$$f(W^s(y, R_i)) \subset \{[f(y), w] \mid w \in W^s(f(x), R_j)\} = W^s(f(y), R_j) \subset S.$$

In other words, $f^{-1}(S) \cap R_i$ is a stable strip of R_i.

Using the equivalent formulation of the Markov conditions given in the remark after the definition of a Markov partition:

$$x \in \mathring{R}_i \cap f^{-1}(\mathring{R}_j) \Rightarrow \begin{cases} f(W^s(x, \mathring{R}_i)) \subset W^s(f(x), \mathring{R}_j), \\ f^{-1}(W^u(f(x), \mathring{R}_j)) \subset W^u(x, \mathring{R}_i), \end{cases}$$

we see that a similar proof gives the result for $\mathring{R}_i$ and $\mathring{R}_j$. □

As usual we have an analogous result with f replaced by f^{-1} and stable strips replaced by unstable strips.

Corollary 10.32. *Let $(a_0, \ldots, a_m)$ be an admissible sequence for A. Then the intersection $\bigcap_{s=0}^m f^{-s}(R_{a_s})$ (resp. $\bigcap_{s=0}^m f^{-s}(\mathring{R}_{a_s})$) is a nonempty stable strip of R_{a_0} (resp. $\mathring{R}_{a_0}$). Likewise, $\bigcap_{s=0}^m f^{m-s}(R_{a_s})$ (resp. $\bigcap_{s=0}^m f^{m-s}(\mathring{R}_{a_s})$) is a nonempty unstable strip of R_{a_m} (resp. $\mathring{R}_{a_m}$).*

Proof. All the assertions are proved the same way: by induction on m using the preceding corollary. We will only prove the first.

For $m = 1$ this is just Corollary 10.31. Suppose that it is true for all admissible sequences $(b_0, \ldots, b_{n-1})$, and let $(a_0, \ldots, a_n)$ be an admissible sequence. Then $(a_1, \ldots, a_n)$ is admissible, so the set $S = \bigcap_{s=1}^n f^{-s+1}(R_{a_s})$ is a stable strip of R_{a_1}, by the induction hypothesis. Applying Corollary 10.31 to R_{a_0} and S, we are done. □

Now let $\underline{a}$ be a point in Σ_A. From the last corollary and the compactness of the R_i's, we see that the sequence $(F_n)_{n \in \mathbb{N}}$, where $F_n = \bigcap_{s=-n}^n f^{-s}(R_{a_s})$, is a decreasing sequence of nonempty compact sets. Therefore, the intersection $\bigcap_{n=1}^\infty F_n$ is nonempty. If x and y were two distinct points in this intersection, we would have

$$d(f^i(x), f^i(y)) \leq \operatorname{diam}(R_{a_i}) < \varepsilon \qquad \text{for all } i \text{ in } \mathbb{Z};$$

but ε is a constant of expansivity for f on Λ, so the intersection is exactly one point.

It makes sense, then, to define a map $\Pi\colon \Sigma_A \to \Lambda$ by

$$\Pi(\underline{a}) = \bigcap_{n=-\infty}^{\infty} f^{-n}(R_{a_n}).$$

We define a set H by $H = \bigcap_{n=-\infty}^\infty f^{-n}(\Lambda - \bigcup_{s=1}^k \partial R_s)$; by Baire's theorem, H is a dense G_δ in Λ. The two following theorems give several useful properties of Π.

Theorem 10.33. *The map Π from Σ_A to Λ is continuous, surjective, and injective on $\Pi^{-1}(H)$. It is a morphism of the local product structure, sending C_i into R_i,* *which makes the following diagram commute*:

$$\begin{array}{ccc} \Sigma_A & \xrightarrow{\sigma} & \Sigma_A \\ \Pi \big\downarrow & & \big\downarrow \Pi \\ \Lambda & \xrightarrow{f} & \Lambda. \end{array}$$

PROOF. Π *is continuous.* Let $\underline{a}$ belong to Σ_A, U be a neighborhood of $x = \Pi(\underline{a})$ in Λ, and $(\underline{b}^n)_{n \in \mathbb{N}}$ be a sequence in Σ_A converging to $\underline{a}$. Let $\{F_n | n \in \mathbb{N}\}$ be the compact sets defined above; there is an n_0 such that for $m > n_0$, $F_m \subset U$. Since the $\underline{b}^n$'s approach $\underline{b}$, there is an n_1, such that

$$\forall i \text{ with } |i| < n_0, \qquad \forall m > n_1 \quad \text{we have} \quad b_i^m = b_i.$$

Therefore, when m is greater than n_1, $\Pi(\underline{b}^m)$ is in F_{n_0}, hence in U. Since U and $(\underline{b}^n)$ were arbitrary, Π is continuous.

Π *is surjective, and injective on* $\Pi^{-1}(H)$. Let x belong to H. For every i, $f^i(x)$ belongs to a unique rectangle R_{b_i} and is in the interior of this rectangle. The sequence $(b_i)_{i \in \mathbb{Z}}$ thus belongs to Σ_A and is the unique preimage of x under Π. Since Π is continuous and Σ_A compact, it follows that $\Pi(\Sigma_A)$ is compact and contains H, hence is all of Λ.

Π *is a morphism of the local product structure.* Clearly $\Pi(C_i) \subset R_i$. Let $\underline{a}$ belong to C_i, $x = \Pi(\underline{a})$; the image of the stable set $W^s(\underline{a}, C_i)$ is contained in $\bigcap_{n=0}^{\infty} f^{-n}(R_{a_n})$. Let y be a point in this intersection; we then have

$$d(f^n(x), f^n(y)) \leq \text{diam } R_{a_n} < \varepsilon, \qquad \text{for all} \quad n \geq 0.$$

In other words, y belongs to $W^s(x, R_i)$, so $\Pi(W^s(\underline{a}, C_i)) \subset W^s(\Pi(\underline{a}), R_i)$. Likewise, we can show that for any $\underline{b}$ in C_i, $\Pi(W^u(\underline{b}, C_i)) \subset W^u(\Pi(\underline{b}), R_i)$; hence $\Pi([\underline{a}, \underline{b}]) = [\Pi(\underline{a}), \Pi(\underline{b})]$.

The commutativity of the diagram being obvious, we are done. □

Theorem 10.34. *Any point of* Λ *has at most* k^2 *preimages under* Π, *where* $k =$ *cardinality of* $\mathbf{R}$.

The proof depends on the following lemma.

Lemma 10.35. *Let* $b_0, \ldots, b_n$ *and* $b'_0, \ldots, b'_n$ *be two admissible sequences with* $b_0 = b'_0$ *and* $b_n = b'_n$. *If* R_{b_i} *and* $R_{b'_i}$ *intersect, for all* i, *then the sequences are equal.*

PROOF. Corollary 10.32 allows us to find points x and y in Λ with $f^i(x)$ in $\mathring{R}_{b_i}$ and $f^i(y)$ in $\mathring{R}_{b'_i}$ for all i, $0 \leq i \leq n$. By hypothesis, for all i, R_{b_i} and $R_{b'_i}$ intersect, so we see that

$$d(f^i(x), f^i(y)) \leq \text{diam } R_{b_i} + \text{diam } R_{b'_i} < 2\rho < \delta.$$

Therefore, $[f^i(x), f^i(y)]$ is defined, $0 \leq i \leq n$.

Now $R_{b_0} = R_{b'_0}$, so $[x, y]$ belongs to $\mathring{R}_{b_0}$. Using the equivalent formulation of the definition of Markov partition, an easy induction gives

$$[f^i(x), f^i(y)] = f^i([x, y]) \in W^s(f^i(x), \mathring{R}_{b_i}), \quad 0 \le i \le n.$$

A similar argument applied to f^{-1}, starting with the observation that $f^n([x, y]) = [f^n(x), f^n(y)] \in \mathring{R}_{b_n'}$, shows that

$$f^i([x, y]) \in W^u(f^i(y), \mathring{R}_{b_i'}), \quad 0 \le i \le n;$$

then, since $\mathring{R}_{b_i} \cap \mathring{R}_{b_i'}$ is nonempty, we must have $b_i = b_i'$.

PROOF OF THEOREM 10.34. Suppose that a point x of Λ had $k^2 + 1$ distinct preimages under Π; $\underline{x}^1, \ldots, \underline{x}^{k^2+1}$ say. We can then find an n so large that all the admissible sequences $(x^i_{-n}, \ldots, x^i_0, \ldots, x^i_n)$ are distinct. Since there are $k^2 + 1$ of them we can find indices i and j such that $x^i_n = x^j_n$ and $x^i_{-n} = x^j_{-n}$. The lemma then gives a contradiction. □

The theorem has an immediate corollary.

Corollary 10.36. $\Pi^{-1}(\text{Per } f|_\Lambda) = \text{Per } \sigma_A$.

Corollary 10.37. *Let $\underline{s}$ and $\underline{t}$ be two preimages of a periodic point x. If $s_i = t_i$ for some i, then $\underline{s} = \underline{t}$.*

PROOF. Let n be a common period of both $\underline{s}$ and $\underline{t}$. The sequences $(s_i, \ldots, s_{i+n})$ and $(t_i, \ldots, t_{i+n})$ satisfy the hypothesis of Lemma 10.35: $s_{i+n} = s_i = t_i = t_{i+n}$ and $f^j(x)$ belongs to $R_{s_j} \cap R_{t_j}$, $i \le j \le i + n$. We must have, then, that $\underline{s} = \underline{t}$. □

We say that a collection of rectangles in *linked* if their intersection is not empty. For each $r \in [k]$ we denote by I_r the set of subsets K of $[k]$ having r elements for which the rectangles $(R_i)_{i \in K}$ are linked. We write elements of I_r as r-tuplets of elements on $[k]$: $(s_1, \ldots, s_r)$ with $s_1 < s_2 < \cdots < s_r$. We denote by $\mathbf{S}_r$ the group of permutations of $[r]$.

For each r in $[k]$ we define two square matrices $A^{(r)}$ and $B^{(r)}$, whose coefficients are indexed by elements of I_r, as follows: for $s = (s_1, \ldots, s_r)$ and $t = (t_1, \ldots, t_r)$, if there is a *unique* permutation μ in $\mathbf{S}_r$ such that $A_{s_i t_{\mu(i)}} = 1$ for all i, then $A^{(r)}_{st} = 1$ and $B^{(r)}_{st} = \text{sign}(\mu) = \pm 1$; otherwise $A^{(r)}_{st} = B^{(r)}_{st} = 0$.

Let L be the largest r in $[k]$ for which I_r is not empty. We are now ready to count the periodic points in Λ.

Proposition 10.38. *For all positive integers p,*

$$N_p(f|_\Lambda) = \sum_{r=1}^{L} (-1)^{r-1} \operatorname{tr}((B^{(r)})^p).$$

PROOF. (a) Denote by Σ_r the set of bi-infinite sequences of elements of I_r and by $\Sigma(A^{(r)})$ the subset of sequences $\underline{\hat{a}} = (\hat{a}_n)_{n \in \mathbf{Z}}$ for which $A^{(r)}_{\hat{a}_n \hat{a}_{n+1}} = 1$, for all

$n \in \mathbb{Z}$. Notice that Σ_r and $\Sigma(A^{(r)})$ are defined from I_r and $A^{(r)}$ just as Σ and Σ_A were defined from $[k]$ and A.

We define a shift $\hat{\sigma}_r$ on Σ_r and $\Sigma(A^{(r)})$ in the obvious way: $(\hat{\sigma}_r(\underline{\hat{a}}))_n = \hat{a}_{n+1}$. Denote by $\mathrm{Per}_p(\Sigma(A^{(r)})$ the set of p-periodic points of $\hat{\sigma}_r$ in $\Sigma(A^{(r)})$.

Given a sequence $\underline{\hat{a}} = (\hat{a}_n)_{n \in \mathbb{Z}}$ in $\Sigma(A^{(r)})$ we write

$$\hat{a}_n = (a_n^1, \ldots, a_n^r) \in I_r.$$

Let μ_n be the unique permutation in $\mathbf{S}_r$ which ensures that $A^{(r)}_{\hat{a}_n \hat{a}_{n+1}} = 1$. Define a sequence $(v_n)_{n \in \mathbb{Z}}$ as follows:

$$\begin{aligned} v_0 &= \mathrm{id} \in \mathbf{S}_r, \\ v_n &= \mu_{n-1} \circ \mu_{n-2} \circ \cdots \circ \mu_0 \qquad \text{if} \quad n > 0, \\ v_n &= \mu_n^{-1} \circ \mu_{n+1}^{-1} \circ \cdots \circ \mu_{-1}^{-1} \qquad \text{if}\, n < 0. \end{aligned}$$

We obtain r elements $\underline{\alpha}^1, \ldots, \underline{\alpha}^r$ of Σ_A by setting

$$\alpha_m^i = a_m^{v_m(i)}, \qquad i \in [r], \qquad m \in \mathbb{Z}.$$

Now, for each m, the rectangles $(R_{a_m^j})_{i \le j \le r}$ are linked; therefore, for all m in $\mathbb{Z}$, and all i and j in $[r]$, we have

$$d(f^m(\Pi(\underline{\alpha}^i)), f^m(\Pi(\underline{\alpha}^j))) < 2 \max_{n \in [k]} \operatorname{diam} R_n < \varepsilon.$$

Since ε is a constant of expansivity for f on Λ, it follows that $\Pi(\underline{\alpha}^i) = \Pi(\underline{\alpha}^j)$, for all i and j in $[r]$.

We can then define a map $\hat{\Pi}_r\colon \Sigma(A^{(r)}) \to \Lambda$ by $\hat{\Pi}_r(\underline{\hat{a}}) = \Pi(\underline{\alpha}^i)$, which does not depend on the choice of i in $[r]$.

(b) Suppose that $\underline{\hat{a}}$ belongs to $\mathrm{Per}_p(\Sigma(A^{(r)}))$. In other words, $\hat{a}_m = \hat{a}_{m+p}$ for all m. We have

$$\begin{aligned} f^p(\hat{\Pi}_r(\underline{\hat{a}})) &= f^p\left(\bigcap_{i \in [r]} \bigcap_{n \in \mathbb{Z}} f^{-n}(R_{a_n^i})\right) \\ &= f^p\left(\bigcap_{n \in \mathbb{Z}} f^{-n}\left(\bigcap_{i \in [r]} R_{a_n^i}\right)\right) \\ &= \bigcap_{n \in \mathbb{Z}} f^{p-n}\left(\bigcap_{i \in [r]} R_{a_n^i}\right) \\ &= \bigcap_{n \in \mathbb{Z}} f^{p-n}\left(\bigcap_{i \in [r]} R_{a_{n-p}^i}\right) \\ &= \hat{\Pi}_r(\underline{\hat{a}}). \end{aligned}$$

Thus we have the inclusion

$$\hat{\Pi}_r(\mathrm{Per}_p(\Sigma(A^{(r)}))) \subset \mathrm{Per}_p(f|_\Lambda).$$

(c) Conversely, suppose the x belongs to $\mathrm{Per}_p(f|_\Lambda)$. It has a finite number of distinct preimages under Π by Theorem 10.34, and they are periodic by

Corollary 10.36. Call them $\underline{\alpha}^1, \ldots, \underline{\alpha}^r$. For all m, the rectangles $R_{\alpha_m^1}, R_{\alpha_m^2}, \ldots, R_{\alpha_m^r}$ are linked and distinct by Corollary 10.37; thus they define an element $\hat{a}_m$ in I_r. We associate to x the sequence $\underline{\hat{a}} = (\hat{a}_m)_{m \in \mathbb{Z}}$ in Σ_r.

Lemma 10.39. *$\underline{\hat{a}}$ belongs to $\Sigma(A^{(r)})$.*

PROOF. It suffices to show that for any given m, the identity is the only permutation in $\mathbf{S}_r$ for which

$$(*) \qquad A_{\alpha_m^i \alpha_{m+1}^{\mu(i)}} = 1, \qquad \forall i \in [r].$$

Suppose not; suppose that $(*)$ also holds for $m = n$ and $\mu = \nu \neq \text{id}$. Let τ be the order of ν in $\mathbf{S}_r$, let j be an element moved by $\nu(\nu(j) \neq j)$, and let q be a common period of the $\underline{\alpha}^j$'s. Consider the two admissible sequences

$$\alpha_n^j \alpha_{n+1}^{\nu(j)} \cdots \alpha_{n+q}^{\nu(j)} \alpha_{n+q+1}^{\nu^2(j)} \cdots \alpha_{n+(\tau-1)q}^{\nu^{\tau-1}(j)} \alpha_{n+(\tau-1)q+1}^{j}$$

and

$$\alpha_n^j \alpha_{n+1}^j \cdots \alpha_{n+q}^j \alpha_{n+q+1}^j \cdots \alpha_{n+(\tau-1)q+1}^j.$$

On the one hand, since we have assumed that $\nu(j) \neq j$. Corollary, 10.37 tells us that $\alpha_{n+1}^{\nu(j)} \neq \alpha_{n+1}^j$. On the other hand, Lemma 10.35 implies the two sequences are equal, and we have a contradiction. □

Now, since x has period p for f, the pth power on the shift leaves invariant the set of preimages of x under Π; from this it follows that $\underline{\hat{a}}$ belongs to $\text{Per}_p(\Sigma(A^{(r)}))$.

(d) Fix $x \in \text{Per}_p(f|_\Lambda)$. We wish to understand the preimages of x under $\hat{\Pi}_r$, for $1 \leq r \leq k$, which belong to $\text{Per}_p(\Sigma(A^{(r)}))$. With $\underline{\alpha}^1, \ldots, \underline{\alpha}^r$ as in (c), let μ be the permutation in $\mathbf{S}_r$ defined by

$$\alpha_p^{\mu(i)} = \alpha_0^i, \qquad \text{for all} \quad i \in [k].$$

The permutation μ is just the permutation of the indices induced by the action of σ^p on the set of preimages of x. Suppose μ is the product of the disjoint cycles $\mu_1, \ldots, \mu_s$ which act on subsets $K_1, \ldots, K_s$ forming a partition of $[r]$.

Let $\hat{b}$ be a preimage of x under $\hat{\Pi}_t$ with $1 \leq t \leq k$. Part (a) shows that $\hat{b}$ gives us t preimages of x under Π, so $t \leq r$. We associate to $\hat{b}$, the subset J of $[r]$ such that the preimages of x under Π given by $\hat{b}$ are just $(\underline{\alpha}^j)_{j \in J}$. If we further require that $\hat{b}$ be p-periodic for $\hat{\sigma}_t$, then it follows from the definition of μ that J in this case is μ invariant, hence a union of certain K_m's. In summary, for each preimage $\hat{b}$ of x under $\hat{\Pi}_t$ which is of period p for $\hat{\sigma}_t$, there is a subset B of $[s]$ such that, setting $J = \bigcup_{m \in B} K_m$, the preimages of x under Π given by $\hat{b}$ are $(\underline{\alpha}^j)_{j \in J}$.

Conversely, for each subset B of $[s]$, letting $J = \bigcup_{m \in B} K_m$ $t = \mathit{card}\ J$, we can, mimicking (c), construct from the preimages $(\underline{\alpha}^j)_{j \in J}$ of x under Π, a preimage $\hat{b}$ of x under $\hat{\Pi}_t$ which is in $\text{Per}_p(\Sigma(A^{(t)}))$.

(e) We are finally able to derive the formula for $N_p(f|_\Lambda)$. Given $\underline{\hat{a}}$ in $\Sigma(A^{(r)})$, let $\underline{\alpha}^1, \ldots, \underline{\alpha}^r$ be the elements of Σ_A constructed in (a). Let ν be the permutation in $\mathbf{S}_r$ which satisfies

$$\alpha_0^i = \alpha_p^{v(i)}, \qquad 1 \le i \le r;$$

notice that sgn v is independent of the numbering of the $\underline{\alpha}^i$'s, since sgn is a conjugacy invariant on $\mathbf{S}_r$. Consider the expression

$$C_p = \sum_{r=1}^{L} (-1)^{r-1} \sum_{\hat{a} \in \operatorname{Per}_p(\Sigma(A^{(r)}))} \operatorname{sgn} v.$$

The proof of Proposition 10.3 also shows that

$$C_p = \sum_{r=1}^{L} (-1)^{r-1} \operatorname{tr}((B^{(r)})^p).$$

On the other hand, from (b) and (c) we conclude that

$$\begin{aligned} C_p &= \sum_{x \in \operatorname{Per}_p(f|_\Lambda)} \sum_{r=1}^{L} \left(\sum_{\hat{\Pi}_r(\hat{a}) = x, \hat{a} \in \operatorname{Per}_p(\Sigma(A^{(r)}))} (-1)^r \operatorname{sgn} v \right) \\ &= - \sum_{x \in \operatorname{Per}_p(f|_\Lambda)} \Phi(x). \end{aligned}$$

The considerations in (d) show that for all x in $\operatorname{Per}_p(f|_\Lambda)$,

$$\begin{aligned} \Phi(x) &= \sum_{B \subset [s]} \left(\prod_{m \in B} (-1)^{\operatorname{card} K_m} \operatorname{sgn} \mu_m \right) \\ &= \left[\prod_{m=1}^{s} (1 + (-1)^{\operatorname{card} K_m} \operatorname{sgn} \mu_m) \right] - 1. \end{aligned}$$

Now the signature of a cycle of length h is just $(-1)^{h+1}$, so

$$\Phi(x) = -1 \qquad \text{for all} \quad x \text{ in } \operatorname{Per}_p(f|_\Lambda).$$

Comparing the two expressions for C_p, we have

$$N_p(f|_\Lambda) = \sum_{r=1}^{L} (-1)^{r-1} \operatorname{tr}((B^{(r)})^p). \qquad \square$$

Theorem 10.40. *The ζ-function of $f|_\Lambda$ is rational.*

PROOF. From Theorem 10.38 and Proposition 10.7 we calculate

$$\zeta_{f|_\Lambda}(t) = \frac{\prod_{r \text{ even}} \det(I - tB^{(r)})}{\prod_{r \text{ odd}} \det(I - tB^{(r)})}. \qquad \square$$

Corollary 10.41. *If $\overline{\operatorname{Per}(f)}$ is hyperbolic, then ζ_f is rational.*

PROOF. From the definition of the ζ-function it is clear that $\zeta_f = \zeta_{f|\overline{\operatorname{Per}(f)}}$. By Proposition 8.11. $\overline{\operatorname{Per}(f)}$ has a local product structure, so we can apply the preceding theorem. $\square$

Corollary 10.42. *Let f be a C^r diffeomorphism ($r \geq 1$) of a manifold M. If f satisfys Axiom* A (resp. *$R(f)$ is hyperbolic, resp. $\overline{L(f)}$ is hyperbolic), then ζ_f is rational.*

PROOF. After Definition 8.9 (resp. Proposition 8.6, resp. Proposition 8.7), Corollary 10.40 applies. □

EXERCISE. Let f be a diffeomorphism satisfying Axiom A. Show that $\Omega(f)$ is finite if and only if $\limsup(1/n) \log N_n(f) = 0$.

Commentary

Symbolic dynamics has a long history, which we will not attempt to cover here. Propositions 10.2, 10.3, 10.7, and 10.8 are from [10.4]. Artin and Mazur [10.2] defined the zeta function of a diffeomorphism and studied some of its properties. Smale [1.16] gave the first proof of the rationality of the zeta function for Axiom A, no-cycle diffeomorphisms. Williams [10.15] and Guckenheimer [10.5] finished the job. Their method was based on the Lefschetz formula. Manning [10.9], using Markov partitions, proved the rationality of the zeta function for Axiom A diffeomorphisms without assuming the no-cycle condition.

Our treatment of Markov partitions is a rewriting of sections of Bowen's book [1.2], which also treats them. We have added some details and formulated the theorems for arbitrary hyperbolic sets with local product structure, not just a single basic set. The proofs are the same and we gain a little in generality, which turns out to be useful. For example, if $f: M \to M$ is an Anosov diffeomorphism, there is a Markov partition even though we do not know that $M = \overline{\text{Per}(f)}$. The systematic introduction of Markov partitions is due to Sinai [10.12], [10.13], [10.14] for Anosov diffeomorphisms and was extended by Bowen [10.3] to Axiom A diffeomorphisms. These Markov partitions are the principal tool for analyzing the qualitative behavior of Axiom A systems. There is a remarkable collection of theorems due essentially to Sinai, Bowen, and Ruslle, see [1.2]. Adler and Weiss [10.1] constructed Markov Partitions for linear hyperbolic automorphism of the two-torus. Our example on T^2 can also be found in Sinai [10.12].

Our proof of (10.32) and (10.33) is made to be parallel to the analysis of the horseshoe in Chapter 4. That analysis could serve as a model for this chapter. Theorem 10.34 was first proved by Bowen [10.3]. Lemma 10.35 is a great improvement in the proof; it was communicated to Bowen by Brian Marcus. We took it from [1.19], which is an excellent survey.

The idea of proving the rationality of the zeta function by using the matrices $B^{(r)}$ was sketched for me by Bowen in 1975; it is very close to Manning's proof. Lebasque and Yoccoz make it coherent here. They have concocted the clearest proof I know of the rationality of zeta functions. Fried [10.8] gives a topological proof relying an isolating blocks and the Leftschetz formula.

The rationality of the zeta function has been used by many authors to obtain qualitative and topological properties of Axiom A systems, see for example [10.6], [10.7], [10.10], and [10.11], or [1.21].

References

[10.1] Adler, R. and Weiss. B., Similarity of automorphisms of the torus, *Mem. Amer. Math. Soc.* **98** (1970).

[10.2] Artin, M. and Mazur, B., On periodic points, *Ann. of Math.* **81** (1965), 82.

[10.3] Bowen, R., Markov partitions and minimal sets for Axiom A diffeomorphisms, *Amer. J. Math.* **92** (1970), 907.

[10.4] Bowen, R. and Landford, O. III. Zeta functions of restrictions of the shift map, in *Global Analysis*, Vol XIV (Proceedings of Symposia in Pure Mathematics), American Mathematical Society, Providence, R.I., 1970, p. 43.

[10.5] Guckenheimer, J., Axiom A + no cycles $\Rightarrow$ $\zeta_f(t)$ rational, *Bull. Amer. Math. Soc.* **76** (1970), 592.

[10.6] Franks, J., Morse inequalities for zeta functions, *Ann. of Math.* **102** (1975), 143.

[10.7] Franks, J., A reduced zeta function for diffeomorphisms, *Ann. of Math.* **100** (1978).

[10.8] Fried, D., Rationality for Isolated Expansive Sets, Preprint, 1983.

[10.9] Manning, A., Axiom A diffeomorphisms have rational zeta functions, *Bull. Amer. Math. Soc.* **3** (1971), 215.

[10.10] Manning, A., There are no new Anosov diffeomorphisms on tori, *Amer. J. Math.* **96** (1974), 422.

[10.11] Shub, M. and Williams, R., Entropy and stability, *Topology* **14** (1975), 329.

[10.12] Sinai, J., Markov partitions and C diffeomorphisms, *Functional Anal. Appl.* **2** (1968), 64.

[10.13] Sinai, J., Construction of Markov partitions, *Functional Anal. Appl.* **2** (1968), 70.

[10.14] Sinai, J., Gibbs measures in ergodic theory, *Russian Math. Surveys* **166** (1972), 21.

[10.15] Williams, R., The zeta function of an attractor, *Conference on the Topology of Manifolds* (Michigan State University, East Lansing, Michigan), Prindle-Weber and Schmidt, Boston, Mass., 1968, p. 155.

List of Notation

Index